Strengthening the Climate–Nature–Food Nexus

V. Ratna Reddy · Dil B. Rahut · Tetsushi Sonobe ·
Anurag Krishna Vippala

Strengthening the Climate–Nature–Food Nexus

Evidence and Policy Lessons from Asia

Springer

V. Ratna Reddy
Livelihoods and Natural Resources
Management Institute
Hyderabad, Telangana, India

Tetsushi Sonobe
National Graduate Institute for Policy
Studies
Tokyo, Japan

Dil B. Rahut
Asian Development Bank Institute
Tokyo, Japan

Anurag Krishna Vippala
Livelihoods and Natural Resources
Management Institute
Hyderabad, Telangana, India

ISBN 978-3-032-29422-7 ISBN 978-3-032-29423-4 (eBook)
https://doi.org/10.1007/978-3-032-29423-4

This Springer imprint is published by the registered company Springer Nature Switzerland AG
The registered company address is: Gewerbestrasse 11, 6330 Cham, Switzerland

Foreword by Dr. Bram Govaerts

If conservation of natural resources goes wrong, nothing else will go right. If agriculture goes wrong, nothing else will have a chance to go right.

—*Dr. M. S. Swaminathan*

Food systems depend on healthy soils, water, biodiversity, and stable climate systems. Climate change, the health of natural ecosystems, and food security are therefore deeply interconnected, and actions affecting one inevitably shape the other, as Dr. M. S. Swaminathan reminded us. Recognizing these linkages is essential for building resilient food systems and achieving sustainable development.

Despite growing recognition of these linkages, empirical evidence on the climate–nature–food (C–N–F) nexus remains limited.

This volume has ambitiously endeavored to address that gap by examining how these systems interact across Asia. Drawing on more than three decades of spatial and temporal data across four Asian sub-regions, the authors analyze the relationships among climate variability, natural capital, and food systems across diverse agro-climatic and socioeconomic contexts. Country case studies further illustrate how these dynamics unfold under different development pathways and policy environments.

The authors highlight the growing tensions between expanding food production and the degradation of natural ecosystems. It shows how the loss of natural capital and biodiversity can intensify climate risks and weaken the resilience of food systems. By introducing an integrated monitoring index, the authors also provide a practical tool for tracking how the nexus evolves in time.

The analysis shows that the climate–nature–food nexus is evolving in an increasingly imbalanced direction in many contexts. Efforts to address hunger and poverty often prioritize short-term gains in food production while overlooking the health of natural systems. Over time, however, degraded ecosystems undermine food security and increase vulnerability to climate risks. The country case studies demonstrate that these dynamics occur across diverse economic contexts and reflect policy choices and investment priorities. At the same time, the study highlights the need for stronger evidence and integrated monitoring systems. It makes a compelling case for continued research and greater investment in data, assessment, and policies that support resilient and sustainable food systems.

The volume also highlights the potential of nature-based solutions to strengthen resilience across climate, nature, and food systems. Yet such approaches remain insufficiently integrated into policy frameworks in many Asian countries and are often implemented in isolation rather than through coordinated strategies.

CIMMYT's long-standing agrifood systems research in Asia focuses on understanding these trade-offs and advancing solutions that strengthen smallholder livelihoods while safeguarding natural resources. This volume provides timely insights into the climate–nature–food nexus and helps lay the foundation for more integrated approaches to sustainable development. For institutions working to advance food security and climate resilience, including CIMMYT, the publication delivers a clear message: resilient food systems and effective climate action depend on healthy, well-managed natural capital.

Dr. Bram Govaerts
Director General
CIMMYT
El Batán, Texcoco
Estado de México, México

Foreword by Dr. Yvonne Pinto

Food systems sit at the center of the world's most consequential policy dilemmas. They must deliver affordable, nutritious diets, and stable rural livelihoods, while operating under mounting climate volatility and accelerating degradation of land, water, and biodiversity. The past decades have produced extensive research on food security and on climate risks to agriculture, and there is a parallel body of work on the conservation of ecosystems and natural capital. Yet, these domains are still too often treated as adjacent problems rather than a coupled system. In practice, strategies are frequently designed and implemented in silos: climate adaptation plans that underweight ecological integrity, conservation efforts that insufficiently engage with food and livelihoods, and agricultural productivity agendas that do not fully internalize climate and nature constraints. The result is a persistent gap between what we understand conceptually about interdependence and what is measured, governed, and financed on the ground.

This book takes that gap seriously. Its central proposition is straightforward: climate, nature, and food production are not three separate policy arenas but mutually reinforcing systems whose interactions shape development outcomes. If this interdependence is ignored, interventions can generate unintended trade-offs and fragility, even when each intervention appears well-designed within its own sector. Conversely, when policy choices recognize the joint dynamics of these systems, it becomes possible to protect environmental assets while sustaining productivity, resilience, and growth. The contribution of this volume is to move beyond general assertions of interconnectedness and to test, using evidence, whether such a climate–nature–food nexus can be observed and characterized at scale.

The authors examine these relationships across Pan-Asia using a long historical lens. Drawing on macro-level data over a 30-year period, the analysis documents how patterns of economic transformation, shifts in resource use, and evolving environmental conditions have co-moved, and how these co-movements differ across countries. The periodization is particularly instructive. The 1990s capture an era in which growth and production objectives often dominated policy, with limited attention to ecological limits. The subsequent decade reflects the payoffs of investment and intensification in many settings, alongside a gradual recognition of environmental stress. By the 2010s, the cumulative costs of degradation and the increasing salience of climate variability became harder to treat as distant externalities, contributing to a more explicit sustainability discourse and, in some cases, policy correction. Taken together, the three decades provide both analytical depth and a policy-relevant narrative of how development pathways interact with environmental constraints.

A distinctive feature of the book is its effort to translate an inherently complex system into a measurable construct. Index-building is never a final answer; it is a disciplined first step that forces clarity about concepts, variables, and implicit trade-offs. The framework offered here invites refinement through better data, higher spatial resolution, and deeper causal testing, but it already serves an important function: it makes the nexus legible to decision-makers who must allocate budgets, set priorities, and justify choices under uncertainty.

The country-focused discussion complements the regional analysis by illustrating how similar dynamics can produce different outcomes depending on initial conditions, institutions, and policy choices. Rather than implying a single "optimal" blueprint, the book highlights that the nexus manifests through multiple pathways, and that effective management requires balancing, sequencing, and adaptation over time. A key message is that sustainability is not synonymous with restrictive conservation, nor is growth synonymous with resource depletion. Durable progress is more often achieved through pragmatic integration: policies that maintain the regenerative capacity of ecosystems, reduce exposure to climate shocks, and keep food systems productive and inclusive.

This volume should be read as an invitation to a different kind of policy conversation. It argues for a shift from single-objective optimization to systems-oriented decision-making, supported by indicators that can be monitored and debated. If the approach is extended and strengthened, it has the potential to inform how countries design climate resilience strategies, target investments in natural capital, and structure food system transformation in ways that are empirically grounded, fiscally realistic, and politically implementable. In that sense, this book is not the conclusion of a research agenda, but the opening of one: a foundation on which more granular analysis, better measurement, and more integrated action can be built.

Dr. Yvonne Pinto
Director General
International Rice Research Institute
Pili Drive, Los Baños
Laguna, Philippines

Preface

As the global south continues to face developmental challenges such as food security, economic stability, and infrastructure growth, the world at large faces a looming climate challenge. While there is a deep understanding of the vulnerability of food production systems to climate and environmental degradation, the inter-relatedness, and more importantly, the interconnectedness of these three systems—Climate, Nature, and Food—is significantly under-represented in the assessment of food security, and climate vulnerability. This is also evident in the mechanisms of interventions being implemented across the world: at best, the relationship between climate and food production systems is taken into consideration when designing climate resilience interventions. The success of these interventions varies for many reasons. Though there is sufficient theoretical understanding of the interdependent and interrelated nature of these three systems, there is insufficient evidence to support the theoretical basis. This book is one of the first attempts to fill this gap and answer the research question: whether there is a nexus between climate, nature, and food systems.

This book examines the nexus between climate, natural, and food production systems at the Pan-Asia level. By examining macroscopic data across Asia, the authors present compelling evidence of this nexus. The analysis also examines the different ways the nexus can manifest, given the varying baseline circumstances of each system. Finally, the authors attempt to approximate a quantitative measure for the nexus, building an index to measure the health of the C-N-F nexus. The book analyses a 30-year time horizon, significant not only for the depth of the dataset but also for what these 30 years represent. From the 1990s through 2020, the dataset captures 3 phases of development and environmental policy. The 1990–2000 period captures the era of aggressive growth policy both in terms of development and food production systems. The 2000–2010 period captures the era when this policy starts to pay off, as countries see the benefits of the investment and resource utilization initiated in the 90s. The 2010–2020 period captures the beginning of some of the ill effects of the policies adopted in the 90s, when natural resource degradation is at its highest and climate variability is most significant, ushering in an era of more sustainable policymaking in parts of the global south.

This lays the foundation for future research on refining the index, leveraging more granular data and variables. The authors also assess the state of the nexus for various countries in Asia, combining quantitative and qualitative assessment to postulate on how the nexus has been managed or needs to be managed in each of the countries viz., Japan, Vietnam, Nepal, and Uzbekistan represent each of the four regions of Asia. The book provides a foundation for assessing how the C-N-F nexus manifests, the challenges of a poorly managed nexus, and the benefits of a well-managed nexus. Most importantly, this approach validates a very important aspect of sustainable development. It recognizes that aggressive conservation is not a desirable approach to effectively managing the nexus. Rather, countries that can balance how these systems interact with each other and keep all three systems in good health achieve the best outcomes. This is especially important in guiding how the global south can navigate the upcoming climate and developmental challenges. Subsequently, it also serves to guide the global north in validating past policy choices while warning against aggressive policy shifts now.

During the process of writing the book, the authors benefited from interactions with various researchers at different stages. Specifically, the authors benefited from insightful discussions with Dr. Raja Rajendra Timilsina, Prof. Bhagirath Behara, and Dr. T. Chiranjeevi during the writing process of the book. Our sincere thanks are due to them. The research for the book was made possible with the support of the Asian Development Bank Institute, Tokyo, Japan. The authors gratefully acknowledge the ADBI's support for conducting this study. We also thank Mami Nomoto and Chiisye Toki for their support in carrying out the study. The views expressed in this book are those of the authors and do not necessarily reflect the views of the institution. The usual disclaimer applies.

Hyderabad, India — V. Ratna Reddy
Tokyo, Japan — Dil B. Rahut
Tokyo, Japan — Tetsushi Sonobe
Hyderabad, India — Anurag Krishna Vippala

Competing Interests The authors have no competing interests to declare that are relevant to the content of this manuscript.

Contents

Part I
Problem in Perspective

Chapter 1
Introduction

1 Background

Food production is inextricably linked to climate and nature due to its direct dependence on the quality of natural resources, such as land and water. Climatic conditions influence the existence, type, extent, and quality of the natural resource base. Nature loss exacerbates climate risks, and climate change in turn contributes to the loss of nature. These interlinkages between nature and climate are most conspicuous in the agriculture, forestry, livestock, and fisheries sectors. The stability and resilience of natural capital are essential for sustaining food production and ensuring food security. While food security and self-sufficiency have been a primary policy objectives, especially among the developing countries, degradation of the natural resource base to achieve these objectives is proving to be a major constraint in enhancing productivity and production (Reddy and Rahut 2025). This stems from the approach adopted to meet the food demands. That is, intensive farm practices adopted over the past four to five decades place a great burden on natural resources to improve output from the limited land bank. Though these practices have helped achieve food self-sufficiency in most Asian countries, the associated impacts or externalities associated with the intensive farm practices are now coming to light. Data establishing the impacts on human and livestock health through contamination of water, soil, food, fodder, etc. is being reported widely (Reddy and Rahut 2025). The degradation of natural capital is further exacerbated by climate change. With the expected temperature rise of 1.5–3 °C, the revenue loss from agriculture in Asia could be in the range of 12–28% of GDP, respectively (IPCC 2023; Mendelsohn 2014). It is estimated that the yields of staple food crops, such as rice, would decline substantially due to temperature rise (Sekhar 2018). These estimates are partial as they consider only the yield impact of climate change and do not include the nature losses associated with unsustainable management practices along with climate risks. The joint impacts of climate and nature loss on food systems would be substantially higher.

V. R. Reddy et al., *Strengthening the Climate–Nature–Food Nexus*,
https://doi.org/10.1007/978-3-032-29423-4_1

The intricate linkages between climate, nature, and food (C-N-F) are compounding our assessment and understanding of global food systems. Often, climate–food linkages are studied without considering nature–climate or nature–food linkages. Factors such as technology, research and development, and policy (which promotes agricultural intensification) contribute to maintaining the food supply in line with demand, without considering the externalities associated with farm intensification. In the absence of a clear understanding of the linkages and their interactions, some argue that food security (production) is no longer a concern, and the main concern now is nutrition security. While this is rightly attributed to demographic stabilization, it fails to acknowledge the increasing risk and uncertainty in food production due to the degradation of natural resources, coupled with climate risks. The severity of food insecurity is well reflected in some countries, particularly in the aftermath of the COVID-19 outbreak and the current global geopolitical climate. Complacency about food security could lead to the rapid erosion of gains made in food production, resulting in increased hunger and poverty across developing countries. This, however, doesn't undermine the importance of nutritional security (access to quality food). Understanding the linkages and trade-offs among climate, nature, and food is critical for ensuring food and nutrition security.

However, these important linkages are seemingly even *less* understood at the policy level, as evidenced by policies that are formulated and implemented in silos, with little or no synergy across the three areas. Hitherto, the policy focus has been on climate change impacts, i.e., adaptation and mitigation. While the resilience of the ecosystem towards climate change impacts is well recognized, the importance of nature or natural capital in ensuring or enhancing resilience is less emphasized. As a result, investments (both public and private) are focused on adaptation and mitigation. While mitigation investments made in the right areas could strengthen natural capital, they seldom incorporate an understanding and assessment of the linkages between climate and nature. Currently, such an understanding is limited to international agencies (for details, see F4B 2021; Portner et al. 2021; Dasgupta 2021). It is essential to ensure a mutually supportive and habitable climate and biodiversity, allowing sustainable and equitable benefits to flow to the community. In the absence of such understanding and action, climate risks linked to nature are not only spreading widely but also intensifying. These risks are both physical and transitional (F4B 2021).

Physical risks are associated with climate impacts and nature loss, which are detrimental to the functioning of ecosystems and their services. Such nature losses impact production processes and the livelihoods of communities, households, and business organizations. These impacts could manifest as soil degradation, increased pest incidence, reduced pollination, decreased air quality, and reduced precipitation, among others. This would have financial implications at various levels for minimizing losses. Transition losses are associated with policy, economic, and social aspects. Policy changes in favor of environmentally friendly activities and processes may impose restrictions and taxes/incentives (such as the "polluter pays" principle or payments for environmental services), which may result in economic losses or gains. This may result in economic losses for some businesses and render certain technologies obsolete, resulting in sunk costs. Losses associated with social aspects include changing consumer preferences, such as a shift to organic, zero-carbon, or net nature-positive products.

The combined impacts of climate change and nature loss exacerbate as policy and investment priorities are limited to addressing climate change impacts. As nature losses continue to escalate, they begin pushing up the cascading effects and costs of compounding climate–nature risks. Businesses are experiencing the economic losses associated with these compounding risks. Investments in climate–nature risk mitigation are economically rational and provide assured returns. Climate change impacts are much severe on agriculture, forestry, and fisheries, along with large built infrastructure for utilities. Besides, nature losses exacerbate the climate risks to agriculture, forestry, and fisheries. The food and beverage, construction, and pharmaceutical industries are the second most exposed to these risks (F4B 2021). Apart from declining soil quality, water stress, poor water quality, high exposure of infrastructure, and a declining availability of raw materials, diminishing disease control impacts human capital, labor availability, and quality, affecting business operations. These combined risks incentivize businesses to enhance their capital allocations towards mitigating or minimizing climate–nature risks. Choosing climate-only or nature-alone options would increase the risks associated with transition. The benefits are expected to be limited if climate–nature investments are made early, before nature losses become irreversible or require long gestation periods (F4B 2021).

Nature-based solutions (NBS) are observed to be the most suitable for addressing the combined adverse climate–nature impacts at the individual and commercial levels (F4B 2021). NBS is defined as: *"the use of natural processes or elements to improve ecosystem functions of environments and landscapes affected by agricultural practices, and to enhance livelihoods and other social and cultural functions, over various temporal and spatial scales."* (Cohen et. al. 2016, p. 5). Unabated climate change impacts exceed the assimilative capacity of ecological and socioecological systems, necessitating reinforcement to cope with climate risks. NBS enhances biodiversity and ecosystem services that reinforce each other. NBS also provides several co-benefits that can be monetized and hence make a good economic rationale. Though hardcore engineering solutions have adverse impacts on nature, they complement NBS solutions in some cases. Similar to the case of combined climate–nature impacts, as well as the assimilative capacity of local systems, the effectiveness of NBS also varies across locations or regions. Designing NBS thus requires an understanding of the extent and intensity of the C-N-F nexus across agro-climatic conditions. Additionally, the direction of the nexus trajectory is also crucial in addressing future developments.

1.1 Missing Nature for Climate

Despite the scientific and logical linkages between nature, climate, and their combined impacts on food systems, the C-N-F nexus attracted research and policy attention only in the late 2010s. It is yet to receive its due attention at the policy and research levels. In the absence of a clear understanding of the combined climate–nature risks, the focus so far has been on assessing the climate impacts on food systems. Even in the case of the agriculture sector (primary food systems), the importance of the natural resource base and the need for its efficient and sustainable management received little attention. The declining viability of the farm sector is often attributed to climate change, which is characterized by changes in precipitation, intensity, rainfall distribution, and temperature. However, the most conspicuous impact of climate change on food production is through changes in water resources. It is expected that rainfall will further decline in the sub-tropics with more frequent extreme events, resulting in aggravating water stress. Further, climate change is likely to alter the soil nutrient balance as well (Toriyama 2005, p. 562). Together, they would adversely impact food production and per-capita food consumption (kcal per capita per day) as well as hunger (population at risk) (Wieb et al. 2019).

While water and soil are also linked to natural habitat and biodiversity, the combined climate–nature risks are much higher. These interactions increase the frequency of floods and droughts, which adversely affect agriculture, forestry, and marine food systems. Extended and repeated droughts could result in desertification, affecting the arable land bank. Extended droughts coupled with frequent flooding could damage forests and impact biodiversity. Destruction of mangroves and marshy lands, along with rising sea temperatures, would reduce the fish catch. Unsustainable farming, forestry, and fishing practices further exacerbate the degradation process. Adopting climate-focused strategies, policies, and investments is neither effective, efficient, nor sustainable, as they aggravate transition risks.

The limited understanding of the combined climate–nature risks is well reflected in climate change research that emphasizes adaptation and mitigation strategies. Even the recent IPCC assessment report (IPCC 2023) emphasizes the importance of adaptation and mitigation strategies. Although it discusses the loss of nature and the importance of biodiversity, it states that unless mitigation and adaptation measures are implemented at the required pace, their costs will rise while their effectiveness declines. It was observed that extreme climate change under high and low emission scenarios would result in an additional 20–36 percent and 11–33 percent of the global population facing hunger by 2050 when compared to the median-level climate change scenario. Therefore, it is crucial to increase food reserves as a strategy to mitigate the adverse effects of climate change on food security. The required reserves could be three times the current reserves in some regions, such as South Asia (Hasegawa et al. 2021). Another study in eastern India suggests that changes in sowing dates could increase wheat yields by 69 percent without impacting rice yields (McDonald et al. 2022). But such an adaptation strategy was only effective in eastern India. This is not a stable strategy from the farmer's point of view, as long as rainfall distribution is changing, as the 1.5 °C target increasingly appears unfeasible, i.e., temperatures are likely to be higher than predicted (Nature Food 2023). Sustainability of such strategies is debatable. In this context, combined climate–nature strategies that strengthen natural capital along with adaptation and mitigation strategies could be more effective and efficient. The combined strategies could be region-specific, depending on the extent of nature loss, exposure to climate change, vulnerability, and resilience.

Hard infrastructure strategies to address climate risks are most often opted for (flood control infrastructure, dams, reservoirs, embankments, etc.), which can cause severe environmental losses in the long run, as river flows are critical for water supply (surface as well as groundwater) and support biodiversity. Obstructing river flows can reduce peak flows and change water chemistry, resulting in numerous unintended negative impacts on habitat and livelihoods. The adverse impacts of nature loss are more severe for vulnerable groups, such as landless, marginal, and small farmers, as they are often the first to lose access to natural resources like water. At the same time, when used strategically, hard infrastructure can complement nature-based solutions and accelerate the restoration of natural capital, particularly in cases of severe ecosystem degradation (F4B 2021). Watershed infrastructures, for instance, can enhance the natural resource base by improving in situ moisture in drought-prone regions, thereby replenishing surface and groundwater resources.

2 Climate–Nature–Food Nexus

Most of the significant and persuasive research in this regard came from international organizations, and consulting firms around the year 2020 (for instance: IPCC 2019, 2023; UNEP 2021; IPBES-IPCC 2021; World Economic Forum 2020, PwC 2021; Finance for Biodiversity (F4B) 2021; Credit Suisse 2021; UKGov 2021; De Netherlandsche Bank 2020; Mackensy and Company 2020). For the first time, since environmental issues have been on the policy agenda, financial institutions and banks have taken such a persuasive approach in addressing the nature–climate risks. These institutions and firms have recently advised business organizations and governments to invest in biodiversity and cautioned about the ineffectiveness of "climate-alone" investments. They also highlight the urgency of these investments.

The interdependencies, trade-offs, and competing demands of climate, nature, and food (C-N-F) constitute a nexus. Managing this nexus ensures the integrity of ecosystem services and sustains natural capital. Before elaborating on the process of nexus, it would come in handy to explain how the terms "climate," "nature," and "food" are used in the present context. "Climate" is mainly used in terms of rainfall and temperature. Extreme events of rainfall and temperatures result in floods and droughts. Here, the terms "climate" and "climate change" are used interchangeably as synonyms. "Nature" includes both use values (natural resources) and non-use values (biodiversity), which include soil, water, forests, habitat, biodiversity, etc. While natural resources like soil, water, forests, biodiversity, etc. (use values of nature) impact food production directly, non-use values (preference, existence, esoteric, etc.) impact indirectly. However, non-use values are not considered in the present analysis, as assessing non-use values is more complex than evaluating the use values. "Food" production systems primarily include agriculture, forestry, and water-based production systems (such as marine aquaculture). All three production systems are critically linked to nature and are affected by climate change.

The interconnectedness and interdependencies among the three components are presented in a simplified manner in Fig. 1. These three components are variable over time, mainly due to their interactions and the externalities associated with their activities. For instance, food was produced in a natural system and climate with no external inputs for centuries. The system was in a state of equilibrium and balance until Anthropocene brought with it a need for organized food production. Sustainable food production systems have exerted limited pressure on natural resources, such as soil and water, over centuries, despite an increasing demand for food. Only during the latter half of the twentieth century did most developing countries adopt farm intensification practices that utilized external chemical inputs. Changing demographics during the latter part of the twentieth century have not only led to intensive farm practices but also resulted in extensive cultivation, extending to marginal lands and forest areas in recent decades. Nature, on the other hand, provides essential resources such as soil, water, and biodiversity, including carbon content, irrigation, soil moisture, habitats, flora, and fauna. The process of intensive (including excessive use of chemical inputs) and extensive practices of food production systems leads to natural resource degradation and nature loss. Agricultural expansion reduces natural infrastructures, such as hedgerows, woodland, and forest, which provide natural flood protection (F4B 2021). Degradation of the natural system undermines its ability to provide ecosystem services. This, coupled with excessive resource extraction, results in declining productivity and production (food and nutrition security).

For instance, excessive use of chemicals results in degradation of soil and water quality, which adversely affects soil carbon content and the nitrification of water. Chemical farming also reduces biodiversity, impacting the pollination process (reduced bee population) and increasing the incidence of pests and diseases (decline in predator population). Overextraction of groundwater not only increases water stress but also accelerates the pace of eutrophication and desertification. The extension of agriculture to marginal lands and forest areas further exacerbates the erosion and degradation of natural capital. The decline and degradation of forests undermine the provision of ecosystem services, including water and air purification, groundwater recharge, and biodiversity, as well as their assimilative capacity to absorb pollutants.

Climate and climate change exert both positive and negative pressures on natural and food systems, and vice versa (Fig. 1). While climate provides the much-needed biotic support (flora and fauna) to the natural system, climate change impacts nature in terms of resource erosion. Extreme events like floods and droughts cause long-term destruction of nature, including soil erosion, forest degradation, and water availability (quantity and quality) issues. Nature acts as a climate-stabilizing or absorbing factor due to its capacity to absorb emissions and improve the microenvironment through forest cover and biodiversity features. Extreme climate events impact food systems directly in terms of land use, input use, land productivity, and food production. Such extreme events have been on the rise in recent years, and droughts are the major contributors to the resource degradation process (WMO 2022). Climate change,

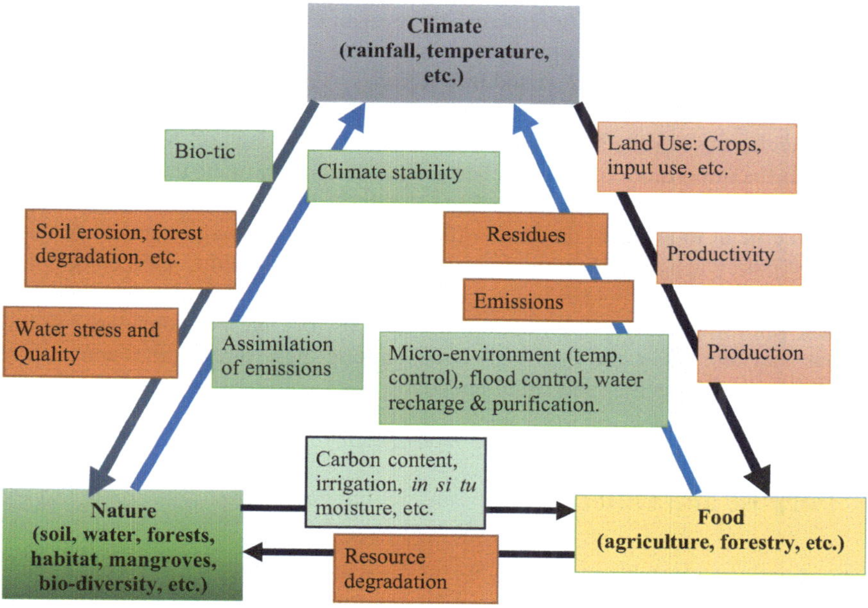

Fig. 1 A simplified framework of climate–nature–food nexus. *Source* Authors

however, may also have positive impacts, such as increased production in some cases or regions. This could be in the form of reduced water stress (increased rainfall, glacial melt, etc.) and improved access to land for crop production (reduced periods of snow cover, increased temperature, etc.).

Food systems also catalyze climate change through emissions and residues associated with agricultural activities (including crop and livestock activities), deforestation, degradation of mangroves, and habitat destruction. And agriculture in all farms (crop, livestock, etc.) contributes substantially to GHG emissions (Lynch 2021). At the same time, food systems (agriculture, livestock, forestry, marine, etc.) contribute positively to the climate in terms of micro-environmental benefits, flood control, groundwater recharge, and water purification (Reddy and Rahut 2023). Thus, the interdependencies among C-N-F components are clear and are becoming increasingly complex over time, with the growing demands on each of the components. A healthy natural environment and its associated ecosystem services are fundamental to the survival of the food system. Degradation of non-crop areas, such as forests and grazing lands (overgrazing), coupled with unsustainable farming practices, erodes topsoil and reduces soil biodiversity. This would result in depletion of organic matter, worsening bioremediation, soil fertility, and water storage. For instance, soil without earthworms can be up to 90% less effective at absorbing water. Similarly, nature loss due to the excessive use of pesticides and herbicides, coupled with climate change (including temperature and rainfall patterns), threatens pollination services (F4B 2021). The current farm practices are causing the faster erosion of resources than the natural system can withstand or replenish on its own. Unless the pace of degradation is checked and resources are replenished continuously, the future of global food and nutrition security remains fragile and uncertain.

Though Fig. 1 is not exhaustive enough to reflect all the externalities, linkages, and interdependencies, it reflects the nature of climate–nature–food (C-N-F) nexus. This highlights the importance of a comprehensive understanding of the three components and their interdependencies for achieving sustained global food and nutrition security. It is crucial to evaluate the trade-offs among the three components to provide informed policy insights. Hitherto, siloed approaches to managing each component independently have not helped in designing effective policies. It may be noted that non-use values are not represented in the figure, as they are not considered in the present analysis. Given the interconnections among these three components, they must be managed together and simultaneously.

3 Purpose and Objectives

Understanding and establishing the C-N-F nexus is in its early stages, as most research is being conducted in silos, focusing on each component separately. So far, the focus has been on climate and food/agriculture across the globe. Although some studies address the C-N-F nexus, they are primarily at the conceptual level. There are no practical case studies using real-world data that help to better understand the C-N-F nexus. The extent and intensity of the nexus are likely to vary spatio-temporally, as its three components are dynamic in nature. Given the regional or agro-climatic-specific nature of the three components, their interdependencies and impacts vary across regions/countries. Similarly, managing the nexus, namely, NBS and its impacts, varies spatially and temporally; that is, the type and nature of NBS (interventions) vary across the agro-climatic spaces of a country or region.

The broad objective of this book is to establish the presence of the C-N-F nexus and to understand its nature across agro-climatic contexts. Further, the linkages, interdependencies, and trade-offs among the three components are explored. It also aims to assess the status, potential, prospects, and constraints for adopting NBS for managing the nexus across agro-climatic situations. Specific objectives include.

(a) Assess the existence and empirically establish the presence of the C-N-F nexus.
(b) Explore the interdependencies and trade-offs among the nexus components across the regions/sub-regions.
(c) Examine the status of C-N-F across geographies and time points.
(d) To organize an integrated monitoring index that helps tracking the status of the nexus.
(e) Assess the role of nature-based solutions in managing the nexus and identify the possibilities for financing NBS beyond public sources.

The research focuses on Asia and its sub-regions. The rationale for the same is as follows:

(i) Asian countries vary widely in terms of agro-climatic conditions and socioeconomic development.
(ii) Asia is home to the largest number of poor people in the world.
(iii) Asia is characterized by highly degraded natural systems (land, water, forests, etc.) and faces the greatest resource stress, especially land and water (WMO 2022).
(iv) Asia is highly vulnerable to nature loss and climate-related risks, including natural disasters, extreme weather events, crop losses, and livelihood disruptions (Shaw et al. 2022).
(v) Asia can provide valuable policy insights and lessons for understanding and managing the nexus in diverse contexts.

4 Approach and Scope

The objectives of this study are to establish the existence of the nexus, assess the interactions between the factors, compare them over time and space, and draw insights from across agro-climatic and socioeconomic contexts. To enable a clear pathway for these objectives, the actions of industrial activity are considered externalities and are therefore not within the scope of this study. Particularly, industrial activity contributes substantially to the degradation of natural systems and the aggravation of climate risks that result in stressed food systems, however require separate work and are therefore outside the scope of this book.

This book focuses on two aspects: first, it presents a comprehensive and systematic review of existing research on the C-N-F nexus, with a particular focus on Asia; secondly, it provides a quantitative assessment of the linkages, trade-offs, and transitions at the Asia, sub-regional, and country levels. A comprehensive and systematic approach has been adopted to gather and review as many aspects of the nexus as possible. These aspects range from conceptual and theoretical to applied and policy-related issues. The review of literature encompasses research studies, including published, unpublished (including gray literature), and online materials. Published studies and data published (public domain) by international organizations, like ADB, FAO, IRRI, OECD, the World Bank, etc., have been used for empirical assessment. This helps to understand and identify the gaps within the C-N-F nexus across countries in Asia and beyond.

Quantitative assessment using secondary data at the sub-regional and country levels for selected countries has been conducted. Quantitative assessment will be focused at two levels: (a) a preliminary assessment of the C-N-F nexus and its trajectories are carried out for the four Asian sub-regions, as established by the Asian Development Bank Taxonomy, viz., Central Asia, East Asia, South Asia, and Southeast Asia; (b) case study assessment of the C-N-F nexus has been taken up for four selected countries representing each of the four regions (Table 1). These countries

Table 1 Asian sub-regions and countries (selected for the study)

Sub-region	Countries
Central Asia (10)	Afghanistan, Armenia, Azerbaijan, Georgia, Kazakhstan, Kyrgyz Republic, Pakistan, Tajikistan, Turkmenistan, Uzbekistan
East Asia (6)	People's Republic of China; Hong Kong, China; Japan; Republic of Korea; Mongolia; Taipei, China
South Asia (6)	Bangladesh, Bhutan, India, Maldives, Nepal, Sri Lanka
Southeast Asia (10)	Brunei Darussalam, Cambodia, Indonesia, Lao People's Democratic Republic, Malaysia, Myanmar, Philippines, Singapore, Thailand, Vietnam
Asia (excluding Pacific Islands, Oceania)	All the above countries

Source ADB classification. https://www.adb.org/sites/default/files/linked-documents/LD1-Glossary-of-Terms-and-Definitions.pdf

are selected based on the status of the C-N-F nexus, categorized as high, medium, or low, using secondary data. Secondary data has been collected for a number of indicators representing each component of C-N-F, viz., indicators of climate (precipitation, temperature, and humidity), nature (land use, biodiversity, water resources, forest resources, etc.), and food (calorie intake, malnutrition, and animal protein availability).

Secondary data analysis has been conducted over a 31-year period, from 1990 to 2020, using the latest available data. However, in the case of some countries and indicators, published data are not available for all the years, and hence analysis is restricted to the years for which data are available. Trendlines, annual growth rates, and one-to-one relationships are used to establish the C-N-F nexus over the years. Correlation matrices are used to examine the strength and direction of the linkages in Asia and its sub-regions. These indicators are used to construct the integrated C-N-F nexus indices, which range between -1 and $+1$. That is, a negative index indicates nexus is resulting in negative impacts, i.e., natural resource degradation, increasing climate risk, and food insecurity. Not all three (C-N-F) need to be in the negative, but the negative impacts (indicators with negative impacts) may outweigh the positives. The reverse holds true in the case of a positive index, and a "0" indicates a balanced status of the C-N-F.

4.1 Limitations

Data demands are quite high given the complexity of establishing the C-N-F nexus. In the absence of sub-regional and country-level data, this assessment does not include certain aspects associated with industrial development and marine food systems. The assessment is based on the time series analysis, which is more robust than cross-sectional analysis, as the latter is likely to include socio-cultural and political dimensions that are difficult to segregate.

5 About the Book

This research aims to establish and assess the "Climate-Nature-Food" nexus in Asia and its four sub-regions. The research draws on published data from international and national databases, including Our World in Data, FAOSTATS, and the World Bank. The book is organized into ten chapters, divided into three parts. In Part 1, which deals with problems in perspective, two chapters are included. Part 2 (regional, sub-regional, and country perspectives) comprises five chapters, covering Asia and its four sub-regions: Central Asia, East Asia, South Asia, and Southeast Asia, as well as four countries representing each sub-region. Three chapters, including conclusions, are covered in part three (managing the nexus). Following the introductory chapter (this chapter), which covers the background, objectives, and approach to the present research, Chap. 2 reviews the available research on the experience of the C-N-F nexus at the broader level. Chapter 2 also presents some evidence on the nexus at the Asia aggregate level using trends in three dimensions of the nexus over a period of three decades (1990–2020).

A quantitative assessment of the nexus is presented for the four sub-regions, as well as for Asia, in Part 2. While Chap. 3 discusses the overall picture for Asia, the remaining four chapters present quantitative evidence for the four sub-regions. Besides, the country case studies are analyzed in these chapters. Uzbekistan from Central Asia (Chap. 4), Japan from East Asia (Chap. 5), Nepal from South Asia (Chap. 6), and Vietnam from Southeast Asia are covered. A quantitative assessment is carried out using correlation analysis for the 30-year period and three sub-periods of one decade each.

Part 3 deals with managing the nexus. An integrated monitoring index has been developed and presented for Asia, its sub-regions, and case study countries in Chap. 8. The role of nature-based solutions (NBS) in managing the nexus sustainably is examined in Chap. 9. Besides, the financing of NBS is also explored in this chapter. The final chapter (ten) draws on the analysis of the preceding chapters, summarizing the findings and offering some concluding remarks. Important policy implications at the regional and country levels are discussed and present future directions as well.

References

ADB (2020) Climate change in South Asia-strong responses for building a sustainable future. Asian Development Bank, Manila

Cohen-Shacham E, Walters G, Janzen C, Maginnis S (2016) Nature-based Solutions to address global societal Challenges. https://portals.iucn.org/library/sites/library/files/documents/2016-036.pdf

Corsi S, Muminjanov H (2019) Conservation agriculture: training guide for extension agents and farmers in Eastern Europe and Central Asia. http://www.fao.org/3/i7154en/i7154en.pd

Credit Suisse (2021) Unearthing investor action on biodiversity. https://www.esg-data.com/product-page/unearthing-investor-action-on-biodiversity

Dasgupta P (2021) The economics of biodiversity: the dasgupta review. HM Treasury, London. https://assets.publishing.service.gov.uk/media/602e92b2e90e07660f807b47/The_Economics_of_Biodiversity_The_Dasgupta_Review_Full_Report.pdf

De Nederlandsche Bank (2020) Indebted to nature. Exploring biodiversity risks for the Dutch financial sector. https://www.dnb.nl/en/actueel/dn-b/dnbulletin-2020/indebted-to-nature/

F4B (Finance for Bio-diversity) (2021) The climate-nature nexus: implications for the financial sector. https://www.naturefinance.net/wp-content/uploads/2022/09/F4B_Climate-Nature-Nexus_Implications-for-Financial-Sector_20210527.pdf

Finance for Biodiversity (2020) Aligning global finance with nature's needs—a framework for systemic change. https://www.f4b-initia-tive.net/news/f4b-releases-white-paper%3A-a-framework-for-systemic-change

FOLU (2019) Growing better: ten critical transitions to transform food and land use. Food and Land Use Coalition. https://www.foodandlandusecoalition.org/wp-content/uploads/2019/09/FOLU-GrowingBetter-GlobalReport.pdf

Hasegawa T, Sakurai G, Fujimori S et al (2021) Extreme climate events increase risk of global food insecurity and adaptation needs. Nat Food 2:587–595. https://doi.org/10.1038/s43016-021-00335-4

Huang B, Xu Y (2019) Environmental performance in Asia: overview, drivers, and policy implications; ADBI Working Paper 990. Asian Development Bank Institute, Tokyo. https://www.adb.org/publications/environmental-performance-asia-overview-drivers-policy-implications

IPBES and IPCC (2021) IPBES-IPCC co-sponsored workshop report on biodiversity and climate change. https://doi.org/10.5281/zenodo.4782538

IPCC (2019) Special report, climate change and land. https://www.ipcc.ch/srccl/

IPCC (2023) Summary for policymakers. In: Climate change 2023: synthesis report. In: Core Writing Team, Lee H, Romero J (eds) Contribution of Working Groups I, II and III to the Sixth assessment report of the intergovernmental panel on climate change. IPCC, Geneva, Switzerland, pp 1–34. https://doi.org/10.59327/IPCC/AR6-9789291691647.001

Iseman T, Miralles-Wilhelm F (2021) Nature-based solutions in agriculture—the case and pathway for adoption. FAO and The Nature Conservancy, Virginia. https://doi.org/10.4060/cb3141en

IUCN (2020) Guidance for using the IUCN Global Standard for Nature-based Solutions, 1st edn. International Union for Conservation of Nature Gland, Switzerland

IUCN (2016) Resolution 69 on defining nature-based solutions (WCC-2016-Res-069). IUCN Resolutions, Recommendations and Other Decisions 6–10 September 2016. World Conservation Congress Honolulu, Hawai, USA. International Union for Conservation of Nature, https://portals.iucn.org/library/sites/library/files/resrecfiles/WCC_2016_RES_069_EN.pdf

John L, Michelle C, David F, Raymond P (2021) Agriculture's contribution to climate change and role in mitigation is distinct from predominantly fossil CO2-emitting sectors. Front Sustain Food Syst 4. https://doi.org/10.3389/fsufs.2020.518039

Keith W, Cattaneo SRA (2019) Climate change, agriculture and food security: impacts and the potential for adaptation and mitigation. In: Campanhola C, Pandey S (eds) Sustainable food and agriculture: an integrated approach. Food and Agricultural Organisation and Elsevier

Kumar Rajesh S, Kundu S, Kundu B, Binu NK, Shaji M (2021) Chapter 10: Emerging typology and framing of climate-resilient agriculture in South Asia. In: Letcher TM (ed) The impacts of climate change. Elsevier, pp 255–287. https://doi.org/10.1016/B978-0-12-822373-4.00021-5
Letcher TM (ed) (2021) The impacts of climate change: a comprehensive study of physical, biophysical, social, and political issues. https://doi.org/10.1016/C2019-0-04893-5
Matthews J, Matthews N, Simmons E, Vigerstol K (2019) Wellspring: source water resilience and climate adaptation. https://www.nature.org/content/dam/tnc/nature/en/documents/Wellspring_FULL_Report_2019.pdf
McDonald AJ, Keil A et al (2022) Time management governs climate resilience and productivity in the coupled rice–wheat cropping systems of eastern India. Nat Food 3:542–551. https://doi.org/10.1038/s43016-022-00549-0
McKinsey and Company (2020) The ESG premium: new perspectives on value and performance. https://www.mckinsey.com/business-functions/sus-tainability/our-insights/the-esg-premium-new-perspectives-on-value-and-performance
Mendelsohn R (2014) The impact of climate change on agriculture in Asia. J Integr Agric 13(4):660–665. https://doi.org/10.1016/S2095-3119(13)60701-7
Miralles-Wilhelm F (2021) Nature-based solutions in agriculture—Sustainable management and conservation of land, water, and biodiversity. FAO and The Nature Conservancy, Virginia. https://doi.org/10.4060/cb3140en
Nature Food (2023) The food-climate nexus, editorial, vol 4, p 271. https://doi.org/10.1038/s43016-023-00745-6
Pörtner HO, Scholes RJ, Agard J, Archer E, Arneth A, Bai X, Barnes D, Burrows M, Chan L, Cheung WL, Diamond S, Donatti C, Duarte C, Eisenhauer N, Foden W, Gasalla MA, Handa C, Hickler T, Hoegh-Guldberg O, Ichii K, Jacob U, Insarov G, Kiessling W, Leadley P, Leemans R, Levin L, Lim M, Maharaj S, Managi S, Marquet PA, McElwee P, Midgley G, Oberdorff T, Obura D, Osman E, Pandit R, Pascual U, Pires APF, Popp A, Reyes- García V, Sankaran M, Settele J, Shin YJ, Sintayehu DW, Smith P, Steiner N, Strassburg B, Sukumar R, Trisos C, Val AL, Wu J, Aldrian E, Parmesan C, Pichs-Madruga R, Roberts DC, Rogers AD, Díaz S, Fischer M, Hashimoto S, Lavorel S, Wu N, Ngo HT (2021) IPBES-IPCC co-sponsored workshop report on biodiversity and climate change. IPBES and IPCC. https://doi.org/10.5281/zenodo.4782538
PwC (2021) The EU taxonomy and the acceleration of sustainable finance. https://www.pwc.com/jp/en/knowledge/column/taxonomy-and-sustainablefinance.html
Reddy VR, Chiranjeevi T (2020) Reorienting watershed management as a climate change adaptation strategy. In: Encyclopaedia of water: science, technology, and society. John Weily
Reddy VR, Rahut DB (2023) Multi-functionality of rice production systems in Asia: a synoptic review. Asian Development Bank Institute, Tokyo, Japan. https://www.adb.org/publications/multifunctionality-of-rice-production-systems-in-asia-a-synoptic-review
Reddy VR, Rahut DB (2025) The future of smallholder farming in South Asia. Asian Development Bank Institute, Tokyo, Japan
Reddy BG, Reddy VR, Reddy MS (2019) Potential and constraints of zero budgeted natural farming (ZBNF): a study of Andhra Pradesh. Indian J Agric Econ 74(3)
Reddy VR, Saharawa YS, George B (2017) Watershed management in South Asia: a synoptic review. J Hydrol 551(2017):4–13
Sekhar CSC (2018) Climate change and rice economy in Asia: implications for trade policy. FAO, Rome, p 62
Shaw R, Luo Y, Cheong TS, Abdul Halim S, Chaturvedi S, Hashizume M, Insarov GE, Ishikawa Y, Jafari M, Kitoh A, Pulhin J, Singh C, Vasant K, Zhang Z (2022) Impacts, adaptation and vulnerability. In: Pörtner H-O, Roberts DC, Tignor M, Poloczanska ES, Mintenbeck K, Alegría A, Craig M, Langsdorf S, Löschke S, Möller V, Okem A, Rama B (eds) Contribution of working Group II to the sixth assessment report of the intergovernmental panel on climate change. Cambridge University Press, Cambridge, UK and New York, NY, USA, pp 1457–1579. https://doi.org/10.1017/9781009325844.012

Toriyama K, Heong KL, Hardy B (2005) (eds) Rice is life: scientific perspectives for the 21st century. In: Proceedings of the world rice research conference held in Tokyo and Tsukuba, Japan, 4–7, November 2004. International Rice Research Institute, Los Baños, Philippines, and Japan International Research Center for Agricultural Sciences, Tsukuba, Japan. CD-ROM. p 590

UK Government (2021) Environment Secretary to set out plans to restore nature and build back greener from the pandemic. https://www.gov.uk/government/news/environment-secretary-to-set-out-plans-to-restore-nature-and-build-back-greener-from-the-pandemic

UNEP (2021) https://www.unep.org/resources/adaptation-gap-report-2020

WMO (2022) State of the global climate. World Meteorological Organisation. https://library.wmo.int/viewer/56300?medianame=1290_Statement_2021_en_#page=1&viewer=picture&o=download&n=0&q

World Economic Forum (2020) Nature risk rising: why the crisis engulfing nature matters for business and the economy. https://www.wefo-rum.org/reports/nature-risk-rising-why-the-crisis-engulfing-nature-matters-for-business-and-the-economy

Chapter 2
C-N-F Nexus: A Review of Experiences and Evidence

1 Context

Food systems are central to the C-N-F nexus. Although food systems impact climate and nature, they are the worst affected by climate and nature risks. Given the importance of food systems in terms of food security and self-sufficiency, the impacts could be devastating. The impacts of degrading natural resource systems, such as soil, water, and biodiversity, are increasingly evident in most countries. Chemical inputs are extensively used as an intervention to address the declining quality of natural resources, thereby further catalyzing the degradation process. Climate change is also negatively impacting food systems in most regions, particularly in Asia. Climate change is progressing and is now more widely experienced and discussed with increasing concrete evidence. From a food systems perspective, important climate indicators include precipitation and its distribution, temperature and its distribution, humidity, as well as the minimum and maximum values of these factors. For instance, it is predicted that global temperatures are likely to increase by 1.5–3 °C by the end of the century. Extreme events, such as floods, droughts, heatwaves, and intense rainfall, are also predicted to intensify. These changes, particularly increased temperatures, are likely to have an adverse impact on agriculture, especially in tropical Asia. The decline in yield rates is expected/estimated to be in the range of 30–50%. This is likely to impact global food security as well.

However, the available evidence on these impacts is very sparse and mostly based on experimental data from controlled laboratory conditions and modeling. In the case of linkages and the C-N-F nexus, the research is more theoretical than empirical. Crop-based food systems contribute less to emissions compared to livestock-based food systems. Crop-based food systems are critical for food security; hence, food systems are the central theme of the debate. This chapter presents a comprehensive and critical review of research pertaining to the C-N-F nexus across the sub-regions of Asia. This chapter aims to establish the linkages using actual data across Asia and its sub-regions over a period of 31 years, i.e., 1990–2020. Additionally, it examines

V. R. Reddy et al., *Strengthening the Climate–Nature–Food Nexus*,
https://doi.org/10.1007/978-3-032-29423-4_2

the theoretical aspects as well as the practical experience of the nexus based on available studies.

2 Review of Experiences

Climate change is the primary concern for food production systems, including crop, livestock, forestry, and fisheries. Changes in frequency, intensity, and geographical distribution of precipitation, temperature, and relative humidity impact food production systems extensively. Increased intensity and frequency of rainfall and temperature often result in extreme events, such as floods (due to increased precipitation and glacial melt) and drought (due to increased temperatures and desertification). Unseasonal rainfall events not only impact crop productivity but also result in significant post-harvest losses. Relative humidity is crucial for crop growth and production, as it affects water balance and photosynthesis. An increase in relative humidity above the optimum level negatively impacts plant growth and production (Chia and Lim 2022; Sawan 2018). However, the impacts vary from crop to crop. Changes in relative humidity also increase post-harvest losses. In the case of livestock, an increase in relative humidity would result in higher disease incidence and a decline in productivity (Yan et al. 2017).

Climate change impacts on food systems and food security are usually assessed under mean- or median-level changes in climate indicators (precipitation, temperature, humidity, etc.). When extreme events associated with climate change are considered, the impacts on the food systems would be substantially higher. In a study that combined crop modeling and extreme climate events, it was observed that an additional 11–36% of the population is likely to face hunger by 2050 under a once in 100-year extreme event in high- and low-emission scenarios (Hasegawa, et. al. 2021). The frequency of such extreme events is likely to increase over time with the progressive increase in emissions. The regional distribution of extreme events would also vary. For instance, South Asia is expected to be more sensitive to these events; hence, the region's current food reserves need to be tripled to address such high climate risks (Hasegawa et al. 2021). While climate risk management requires simple adaptation practices, such as adjusting sowing dates in specific regions (McDonald et al. 2022), it also necessitates expensive mitigation and adaptation measures in other regions, particularly in the context of extreme climate events.

When temperature rise alone is considered, net revenues from agriculture in Asia are expected to decline by 13 and 28% by 2100 in the 1.5 °C and 3 °C scenarios, respectively (Mendelsohn 2014). Global warming may also have a positive impact on crop yields through the mechanism of carbon fertilization. In the scenario of carbon fertilization associated with a temperature rise, Asian agriculture is likely to gain 3% in net revenues under a 1.5 °C scenario and experience a 12% loss under a 3 °C scenario, although this may not be uniform across countries within Asia (Mendelsohn 2014). This study introduces a new dimension to climate change impacts, specifically the positive externalities associated with global warming that benefit crop production. Considering such externalities is important for assessing the net impacts of climate change. Similarly, agriculture and crop systems are associated with a number of externalities in the form of non-commodity benefits. It is observed that non-commodity benefits outweigh the commodity benefits in the case of rice production systems in Asia (Reddy and Rahut 2023).

In the absence of an understanding and assessment of the non-commodity benefits of crop production systems, it is often argued that food systems contribute 34%, or a third, of the total GHG emissions (Crippa et al. 2021; Xia et al. 2023). These high estimates are mainly due to: (i) considering crop- and livestock-based food systems together and (ii) not considering the non-commodity benefits of crop systems. It was estimated that only 29% of the 34% (meaning only 9.86% of total food system emissions) was due to plant-based production systems, while 57% (meaning 19.38% of all food system emissions) corresponds to animal-based food systems and the remaining being accounted for by other utilizations of the food systems, viz., retail, transport, consumption, fuel production, waste management, industrial processes, and packaging (Xu, et. al. 2021). Within the plant-based food systems, rice is the major contributor (12%) while beef is the major contributor (25%) to the livestock food systems. Beef generates the most GHG emissions per million kilocalories consumed, at 226.5 tons of carbon dioxide equivalent (tCO2e), and has the biggest water footprint, at 9,850 cubic meters (m^3), compared with plant-based GHG emissions at 7 tCO2e and water use at 1,300 m^3 (ADB 2021). Asia, due to its extensive rice production systems, and South America, due to its extensive beef production systems, were the largest emitters of food production-based greenhouse gas emissions. Even with its current consumption pattern in Asia, beef still dominates in terms of GHG emissions, although the higher average consumption of rice per capita in the region makes GHG emissions per capita from rice significant. Rice and wheat also use a lot of water per capita, followed by fish. In terms of land utilization, pork has the highest per capita usage, at 297 m^2, followed by wheat and rice (ADB 2021).

Given the very high contribution of livestock-based production systems, it is argued that a shift from animal-based to plant-based foods could substantially reduce greenhouse gas emissions from direct agricultural production. The shift primarily needs to occur among high-income nations, as it could reduce emissions by 61%, equivalent to approximately 14 years of current global agricultural emissions until natural vegetation matures (Sun et al. 2022). Even a shift from beef to small ruminants would help reduce nitrogen emissions by 2% and greenhouse gas emissions by 5% due to land-use change and lower demand for cropland areas for ruminant feed (Cheng et al. 2022). The output from released cropland could feed up to 525 million people worldwide. Thus, it is critical to understand and establish that crop-based food systems cause the least damage in terms of GHG emissions. Unless crop and livestock-based food systems are assessed separately, we risk sidelining a food system that can significantly contribute to both food and nutrition security, while leaving a much smaller carbon footprint behind. It is imperative to establish solutions that account for these differences in production, transportation, and caloric as well as nutritional value.

While research studies have identified several solutions, there remains a lack of clarity on achieving sustainable food production systems. This is primarily due to an absence of holistic and integrated approaches to addressing climate and food system linkages. The current approaches overlook some fundamental principles of environmental economics, such as externalities associated with climate change risks and mitigation interventions. Frameworks like life-cycle assessment (LCA) and circular economy (CE) are not effectively adopted. Adoption of such frameworks helps provide a comprehensive understanding and design-integrated mitigation strategies. It is argued that land-linked emission reduction strategies may have adverse impacts on food security. For instance, research has shown that afforestation (often simulated in models by imposing carbon prices on land carbon stocks) could have a significant impact on food security compared to non-CO2 emission policies (Fujimori et al. 2022). Afforestation measures reduce the land available for crop production, putting an additional 41.9 million people at risk of hunger by 2050 compared with the baseline scenario. This highlights the need for better coordination in emissions reduction and agricultural market management policies, as well as improved representation of land use and associated greenhouse gas emissions in modeling.

On the other hand, a life-cycle assessment of food production systems helps reduce life-cycle greenhouse gas emissions from staple crops through an integrated biomass pyrolysis and electricity generation system, coupled with commonly applied methane and nitrogen mitigation measures (Xia et al. 2023). Emission reductions would be achieved primarily through carbon sequestration from biochar application to the soil, and fossil fuel displacement by bio-energy produced from pyrolysis. Similarly, GHG emissions from food loss and waste (FLW) are not assessed from a life-cycle approach of "cradle to grave." A study employing the "cradle-to-grave" approach identified nine post-farming sources of FLW emissions that vary by country, region, and food category. Income level, technology availability, and prevailing dietary patterns also affect country and regional FLW emissions (Zhu et al. 2023). Halving FLW generation, halving meat consumption, and enhancing FLW management technologies are the strategies assessed for FLW emission reductions.

In the literature on the food–climate nexus, it is well established that they contribute to each other through mutual interactions. The latest IPCC synthesis report (IPCC 2023) clearly indicates that food systems continue to affect climate change and vice versa. For the first time, it introduced the nature dimension to the interactions and trade-offs. As a result, it postulates that technology alone can no longer provide the solutions. Technological solutions would be more effective when addressing two-dimensional problems, such as climate change and food security. The sustained climate–food nexus over decades has impacted and added the third dimension to the nexus, i.e., climate–nature–food. In this context, the IPCC argues that institutional, financial, political, and behavioral aspects play a crucial role in identifying appropriate measures and solutions. The excessive extraction of natural resources in the process of meeting food demands that often exceed real needs, due to poor distribution and the greed of a few, is degrading the natural systems, which are at the root of the problem (Reddy and Behara 2025). More importantly, diversion of resources away from basic needs and requirements (for a decent living) is putting undue pressure on the natural system.

Nature acts as the basis for food systems and mitigating climate risks. Nature, in the form of biodiversity and ecosystem services, supports and sustains food systems. At the same time, natural capital provides the much-needed resilience against climate risks at different scales (from households to ecosystems). Despite the importance of nature, it has not received its due attention at the research and policy levels. Most of the available studies on nature–climate linkages are of recent origin, and most of them are prepared as background research for the latest IPCC report (2023). These studies focus on biodiversity as a primary indicator of nature and its potential to mitigate climate risks (Dasgupta 2021; IPBES and IPCC 2021). It is argued that investing in and financing nature and nature-based solutions is likely to make climate projects financially viable (F4B 2021). Similar to the case of climate and food linkages, nature is closely linked to climate, and thus all three aspects are interconnected. In the absence of such an integration, the understanding and assessment of interlinkages remain partial, making mitigation measures ineffective.

Deterioration of the nature not only adversely impacts food production but also slows down the process of mitigating climate risks. In some cases, such as tropical forests, the damage could be irreversible. Given that most of the present research is based on modeling using historical trends fostered by scientific experiments in controlled (laboratory) conditions. And most of these studies do not integrate the nature dimension into their predictions. This makes the future trajectories uncertain and tends to go wrong, either way. For instance, it is estimated that under controlled conditions, higher temperatures improve carbon fertilization of soils, resulting in increased net revenues/reduced net revenue losses from crop production by 2100 under 1.50 °C and 30 °C scenarios, respectively. These predictions are based on several assumptions, including the climate sensitivity of crop systems and access to irrigation (Mendelsohn 2014). Both assumptions are critically linked to the status of natural capital. Deterioration of natural ecosystems by 2100 could render crop systems less resilient or more sensitive to climate change and exacerbate water stress.

Similarly, although afforestation intervention is expected to impact food production adversely (Fujimori et al. 2022), it is necessary to account for the positive impacts of forest cover on ecological indicators like biodiversity, carbon sequestering, water recharge, etc. Assessment of the impacts of crop production systems considers only the GHG emissions, while ignoring the multiple non-commodity benefits associated with these systems. Crop production systems are characterized by multi-functionality, as they produce joint products in all their activities using non-allocable inputs, i.e., producing multiple outputs from the same inputs (OECD 2000). More importantly, crop systems produce non-commodity outputs, such as groundwater recharge, water/air purification, and landscape benefits, using the same inputs. These non-commodity outputs originate from land use and crop practices, resulting in externalities that are both positive and negative, intended and unintended. Externalities may also result in socioeconomic benefits or disbenefits at the individual, national, and global levels. However, these externalities are not taken into account when assessing the impact of crop-based food systems on GHG emissions.

Once these externalities are considered, crop production systems might have a net negative impact on emissions.

Although multi-functionality is a subset of ecosystem services, policymakers or researchers in most developing countries are either unaware of or do not acknowledge it. Even farmers are not aware of these services, as the benefits are social/public in nature rather than private. More importantly, the majority of these services don't have markets, i.e., they are not transacted in normal markets (popularly known as market failure). While the economic value of multiple commodity outputs (including joint or by-products) is well considered (from individual to national and global levels) by farmers as well as policymakers, little attention is paid to the non-commodity outputs (services and their externalities) they provide (OECD 2000).

The multi-functionality of crop systems is critical due to its close linkages with the sustainable development goals. Addressing externalities for sustaining ecosystem services is integral to achieving multi-functionality and sustainable development, thereby ensuring the production potential for future generations. This has helped multi-functionality to gain validity in terms of economic rationale, environmental sustainability, scientific credibility, and social objectivity. Social objectivity becomes evident when multi-functionality is adopted in smallholder-dominated agrarian economies, such as those in Asia (Reddy and Rahut 2023). Thus incorporating multi-functionality into economic policy helps addressing three important areas of concern: i) rationalizing and increasing the allocations towards agriculture by incorporating the value of its environmental services; ii) achieve sustainable development goals of food security, poverty alleviation, etc., through mitigation/minimizing the negative impacts of externalities; and iii) achieve socially inclusive sustainable development through improving the viability and resilience of smallholder farmers in the developing countries (FAO 2004). This, in the long run, can change the terms of trade in favor of agriculture.

The preceding review highlights several shortcomings in our understanding of the C-N-F nexus. One of the main reasons for this is that "nature" has recently entered the third dimension and continues to evolve and establish itself. The research focus continues to be two-dimensional, i.e., either "climate–food" or "climate–nature." As a result, the assessment of linkages and trade-offs is not comprehensive. At the methodological level, the present research is primarily based on aggregate or macro-data, and therefore significant micro-realities or impacts are unaccounted for due to geographical variations. Some of the aggregate impact assessments are not alarming (Mendelsohn 2014). However, ignoring the micro-impacts and inferences could prove costly in the long run, particularly in the context of the cascading effects of these strained micro-impacts. Furthermore, analysis and inferences regarding climate–food interactions are based on data obtained from controlled or laboratory experiments. Despite being scientific, the assessments may differ widely from reality or from ground situations.

Very few studies employ comprehensive assessment and evaluation tools, such as "life-cycle assessment," "circular economy," or "nexus." As a result, the assessments fail to integrate the externalities that form the basis for understanding interactions between environmental and natural systems. The study by Mendelsohn (2014) assessed potential gains from crop-based food systems due to externalities, such as carbon fertilization, in association with temperature rise. However, these gains are contingent upon natural attributes, including the widespread availability of sufficient irrigation. Additionally, the multi-functional characteristics of crop systems with substantial externality impacts are largely ignored in the assessments. Including these externalities could make crop-based food systems net contributors to mitigation, rather than net emitters. These are only a few examples, and a comprehensive scoping of the externalities associated with the C-N-F nexus is required. Such externalities (positive and negative) need to be integrated into the assessments.

On the whole, the following broad inferences could be drawn from the review:

(i) Need to consider nature as integral to "climate and food" systems and make the assessments three-dimensional.
(ii) Aggregate data analysis strips away the meaningful context of micro-level changes that have a real impact on society and economies at large.
(iii) These micro-variables continue to be examined in isolation from one another, meaning that too many what-ifs already hinder any analysis.
(iv) Accounting for externalities can provide greater insight into how policy can effectively guide technology to leverage this to meet food needs without degrading the natural system.
(v) Modeling alone cannot be the war cry for climate change management.
(vi) Qualitative analysis, especially when examining micro-realities, proves more valuable than relying entirely on quantitative analysis.

A systematic presentation of the C-N-F nexus and its trajectories over the years across various regions of Asia is presented in the following section.

3 C-N-F Nexus: Regional Trajectories

The number of indicators representing each dimension of the C-N-F nexus is available from global public sources, such as "Our World in Data," "Food and Agricultural Organization (FAO)," and "The World Bank," among others. The most appropriate and important indicators representing each aspect are identified for the purpose of this analysis. All the indicators represent the basic characteristics and are directly associated with each aspect or dimension of the C-N-F nexus. No indirect indicators, such as pollution, emissions, access to water, poverty, or income, are included here. These indicators will be considered in a more detailed analysis later. Details of the selected indicators are presented in Table 1. Data is not available in the required form for some important indicators, such as average daily temperature. Alternative measures, such as annual temperature anomalies, are used instead.

In the case of nature, not many indicators are available, i.e., only forest cover and the Living Planet Index (LPI) are available at the sub-regional level, while the Red List Index (RLI) is available for the Asia–Pacific region only. LPI is based on the Living Planet Database (LPD), which currently holds time series data for almost

Table 1 Selected indicators and their measurement

Aspect/ dimension	Indicator	Measurement
Climate	(a) Average annual precipitation (b) Precipitation anomaly (c) Temperature anomaly (d) Relative humidity	(a) In centimeters/decimeters/meters (b) Deviation from average annual precipitation in mm/cm/decimeters (c) Deviation from the average annual temperature in degrees centigrade (d) In percentage
Nature	(a) Forest cover (b) Living Planet Index (LPI) (c) Red List Index (RLI)	(a) % of area under forest to total area (b) Global biological diversity based on population trends of vertebrate species from around the world (c) Overall extinction risk for species
Food	• Total calorie supply • Extent of malnutrition • Supply of animal protein • Share of animal protein	(a) Kilo calories per capita/day (b) % of the malnourished population (c) Grams per capita per day (d) In percentage of total calorie consumption

Source Authors

42,000 populations of 5,579 mammal, bird, fish, reptile, and amphibian species from around the world. These data are gathered from a variety of sources, including journals, online databases, and government reports. These species population trends are aggregated to produce indices of the state of biodiversity. RLI displays trends in overall extinction risk for species and is used by governments to track their progress towards targets for reducing biodiversity loss. Four indicators are used to represent the food aspect. These indicators reflect hunger, poverty, nutrition, and consumption habits. Data on all the selected indicators are analyzed for 30 years, i.e., 1990 to 2020. It is worth noting that data is available for only 20 years, specifically from 2001 to 2020, in the case of the extent of malnutrition. The analysis is carried out for Asia as a whole and its four regions (Central, East, South, and Southeast Asia).

3.1 C-N-F Nexus: Trends in Asia

Given the nature of the climate, all three indicators exhibit high year-on-year fluctuations over the past 30 years (Fig. 1). Precipitation anomalies reveal higher year-on-year fluctuations compared to temperature anomalies. Relative humidity also showed high fluctuations between the years, although they fluctuated in a narrow range between 70 and 68% over a period of 30 years. When trend lines are drawn, both precipitation and temperature anomalies have shown an increasing trend, while relative humidity has revealed a declining trend. Temperature anomalies demonstrated a relatively sharper trend when compared to precipitation anomalies. This indicates that both precipitation and temperature recorded increases above the mean temperature and rainfall. Precipitation anomalies range between −1.5 and +1.17 cm, while temperature anomalies range between −0.82 and +0.65 °C. It is worth noting that temperature anomalies in recent years (after 2012) have consistently been in the positive range. The temperature trends are commensurate with the IPCC predictions on global warming. But the aggregate temperature anomalies do not suggest any extreme trends.

Biodiversity is a primary indicator that reflects the state of nature. Most recent studies have considered biodiversity as the most important determinant influencing ecosystem services, which impact both climate and food systems. Given that forests are the main habitats of biodiversity in terms of flora and fauna, three important indicators—extent of forest cover, Living Planet Index (biodiversity index), and extent of endangered species (Red List Index)—are considered in the present analysis. All three indicators have shown a declining trend over the past 30 years (Fig. 3). The decline in the share of forest cover in the total geographical area is very moderate over the past 30 years in Asia. This could be due to policy changes aimed at protecting forests in most Asian countries, especially during the 1990s. These policies include restrictions on developmental activities in forest areas, forest

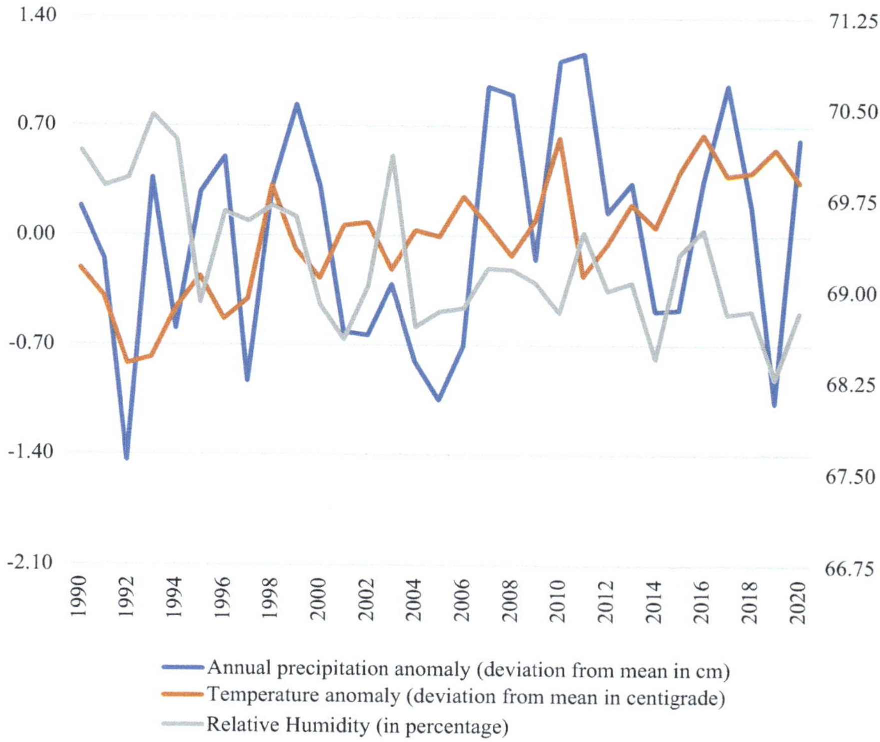

Fig. 1 Trajectories of climate indicators in Asia. *Source* https://ourworldindata.org/climate-change

management and protection through participatory forest management, protection of indigenous communities and their rights, and afforestation interventions, among others. While the area under forest cover remains stable, the quality of forests appears to have declined despite numerous forest protection interventions. This is reflected in the sharp decline in the biodiversity index (LPI) over the years. Over a period of 30 years, LPI in Asia declined from about 91% in 1990 to about 40% in 2020 (Fig. 2). On the other hand, the index of species under risk has declined marginally, which could be attributed again to the policies pertaining to wildlife protection in most Asian countries.

Food is often considered in terms of quantity, such as three meals a day or the calories supplied per capita per day. The quality of food, including its composition (such as carbohydrates and protein) and nutritional value, is rarely considered in climate–food linkages or discussions. While the number of meals and calorie availability reflect the hunger/poverty dimension, they do not indicate the well-being of the people if the food consumed is not balanced and nutritious. Nutritious food is essential for maintaining the health of the human body, and the composition of food plays a crucial role in sustaining the body's health and overall well-being. The composition of food, in terms of the consumption of plant-based and animal-based foods, has implications for natural resource use and the efficiency of food systems. These, in turn, will have implications for environmental and resource degradation through

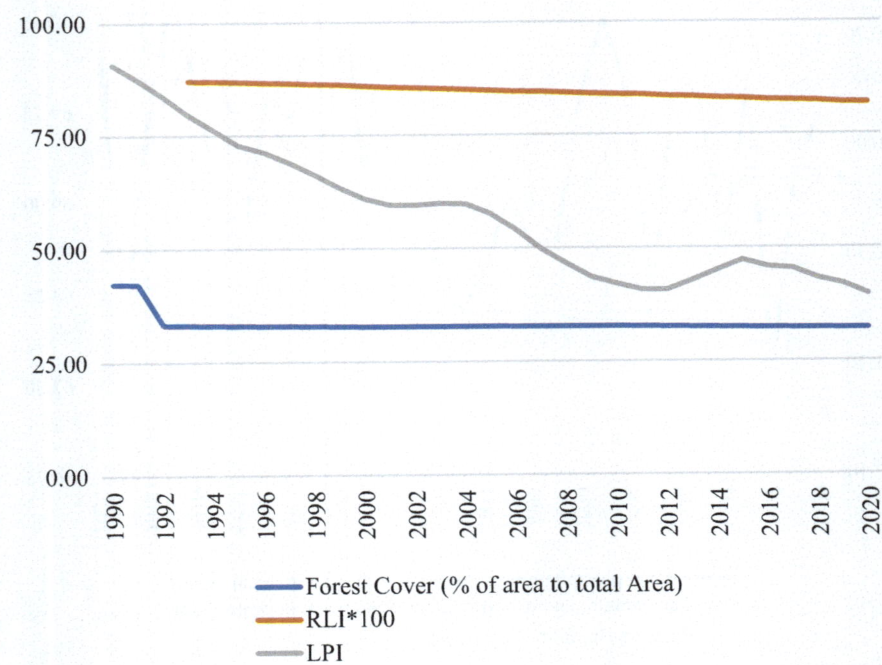

Fig. 2 Trajectories of nature indicators in Asia. *Source* https://ourworldindata.org/climate-change

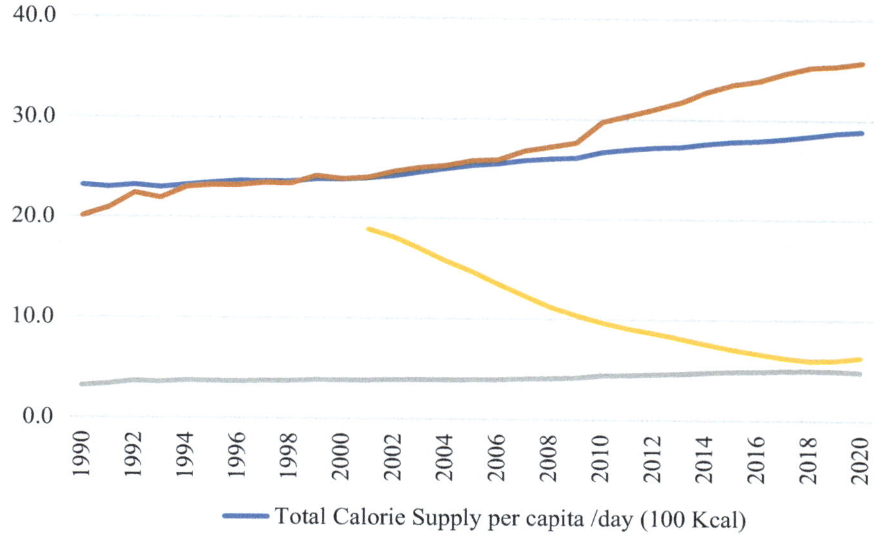

Fig. 3 Trajectories of food indicators in Asia. *Source* https://ourworldindata.org/climate-change

land use and emissions. For instance, animal-based food products consume more resources and produce higher emissions compared to plant-based food products. In fact, a complete shift in food preferences towards plant-based food could address a substantial global emissions burden, especially among developed countries.

Three indicators have been chosen to represent hunger/poverty, nutrition, and the composition of food, namely, total calorie supply, supply of animal protein, and the extent of malnutrition. Besides these, the share of animal protein in the total calorie supply is also considered. All indicators suggest an improved contribution of food systems in Asia (Fig. 3). The total calorie supply has increased from 2300 kcal (the minimum required) to approximately 2900 kcal per capita/day over the 30-year period. The supply of animal protein has also increased from 20 g per capita/day to approximately 36 g per capita/day. This has resulted in a substantial decline in the prevalence of malnutrition in Asia, with the extent of malnutrition decreasing from 18 to 6% of the total population. Despite a substantial increase in the supply of animal protein, its share in the total calorie supply in Asia has increased marginally from 3% to about 5% over a 20-year period, as data on malnutrition is available only from 2001.

In order to visualize the trends in the C-N-F nexus, three important indicators representing each of the three dimensions are presented in Fig. 4. Looking at the trajectories of the three dimensions over the past three decades in Asia indicates that while increasing temperatures characterize the climate, nature is being degraded in terms of biodiversity and forest cover. These two dimensions are expected to reinforce each other negatively in the medium to long run. Climate change (increase in temperature) due to anthropogenic activities is likely to degrade the natural system (forests, biodiversity, etc.), which in turn reduces its assimilative capacity, i.e., absorbing the emissions, resulting in an increased pace of global warming. This vicious cycle of the climate and nature nexus will impact food systems both directly and indirectly. Increased temperatures directly impact the productivity of certain crops, and degradation of nature indirectly affects the food systems through declining ecosystem services, such as soil quality and water availability. However, our assessment in the Asian context doesn't seem to have had a direct impact on food systems, i.e., affecting the availability of food or decline in food production. One reason could be that the negative impacts on food systems are not yet widespread due to technological interventions, especially in crop-based systems.

For instance, green revolution technologies have substantially enhanced land productivity and contributed to global food security. These technologies have also sown the seeds of unsustainable, input-intensive farm practices (excessive use of chemical fertilizers along with water) that have resulted in land and water resource degradation (Reddy 2003, 2005; Reddy et al. 2018). The over-intensification of land resources (utilization of high-cost inputs), coupled with the over-exploitation of groundwater (resulting in water stress) and the expansion of crop cultivation, has enabled high productivity and production levels to meet the increasing food demand. But chemical intensification is impacting farm viability, as input costs are growing at a faster rate than crop value. In fact, some regions, such as Indian Punjab and certain

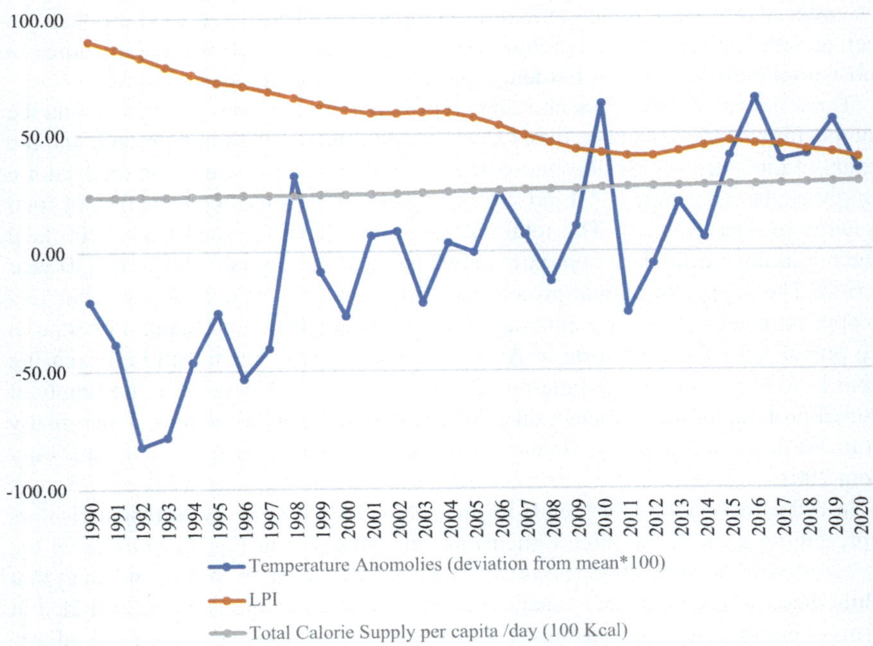

Fig. 4 Climate, nature, and food trajectories in Asia. *Source* https://ourworldindata.org/climate-change

dry regions, are experiencing productivity losses due to soil and water degradation (Reddy et al. 2018). Climate variability or change is further adding to the complexities. Recently, climate change impacts are equally evident in pre- and post-harvest (loss of produce) farm activities, resulting in declining production. The declining viability of farming is driving farmers out of agriculture in several Asian countries (Reddy et al. 2025). Thus, while the impacts of the climate–nature nexus on food systems are not yet explicitly visible, the impacts are implicit in terms of farm viability. Additionally, the direct impacts are currently limited to specific regions and crop systems. The regional assessment of the C-N-F nexus in the following section may provide some insights in this regard.

3.2 C-N-F Nexus: Regional Variations

Given the character of climate and nature, as well as their geographical differences, aggregate indicators at the Asia level could obscure the sub-regional variations. To capture the sub-regional differences, this section presents the trends in selected indicators for the different sub-regions of Asia. For this purpose, four sub-regions within Asia are selected, namely, the Central, East, South, and Southeast Asia regions. These four regions encompass all the major countries in Asia, excluding only the Pacific Islands. Regional variations are observed in the case of all the selected indicators, although the trends over the past three decades do not vary significantly across the regions (Figs. 5 and 6).

In the case of nature indicators, there are wide variations in the percentage area under forest cover across the regions. It ranges from 10% in Central Asia to 50% in Southeast Asia. Over the last three decades, only Southeast Asia has experienced a decline in forest cover, i.e., from 55 to 50%, while the other three regions have reported a marginal increase (Fig. 6a). The regional variations are marginal in the case of RLI and all the four regions have reported a decline in the endangered species (Fig. 6b). Among the four regions, South Asia has the lowest number of endangered species (an index value of 0.73 in 2020), while Central Asia has the highest number (an index value of 0.92). In the case of biodiversity, data are available only for the Asia–Pacific region. This is compared with the global biodiversity index. The Asia–Pacific region has higher biodiversity compared to the rest of the world, and the gap between the two was as high as 30 points in 1990 (Fig. 6c). However, there has been a sharp decline in biodiversity in the Asia–Pacific region over the past 30 years. The decline is much less pronounced in the case of global biodiversity, which has reduced the gap. This indicates that Asia is experiencing greater degradation of biodiversity. Although sub-regional data is not available in this regard, based on the trends in the case of the Asia–Pacific aggregate, the decline could be similar in the sub-regions, with marginal regional differences.

All four sub-regions of Asia have recorded increasing trends in all food indicators, although the magnitude of these indicators varies across the regions. East Asia is at

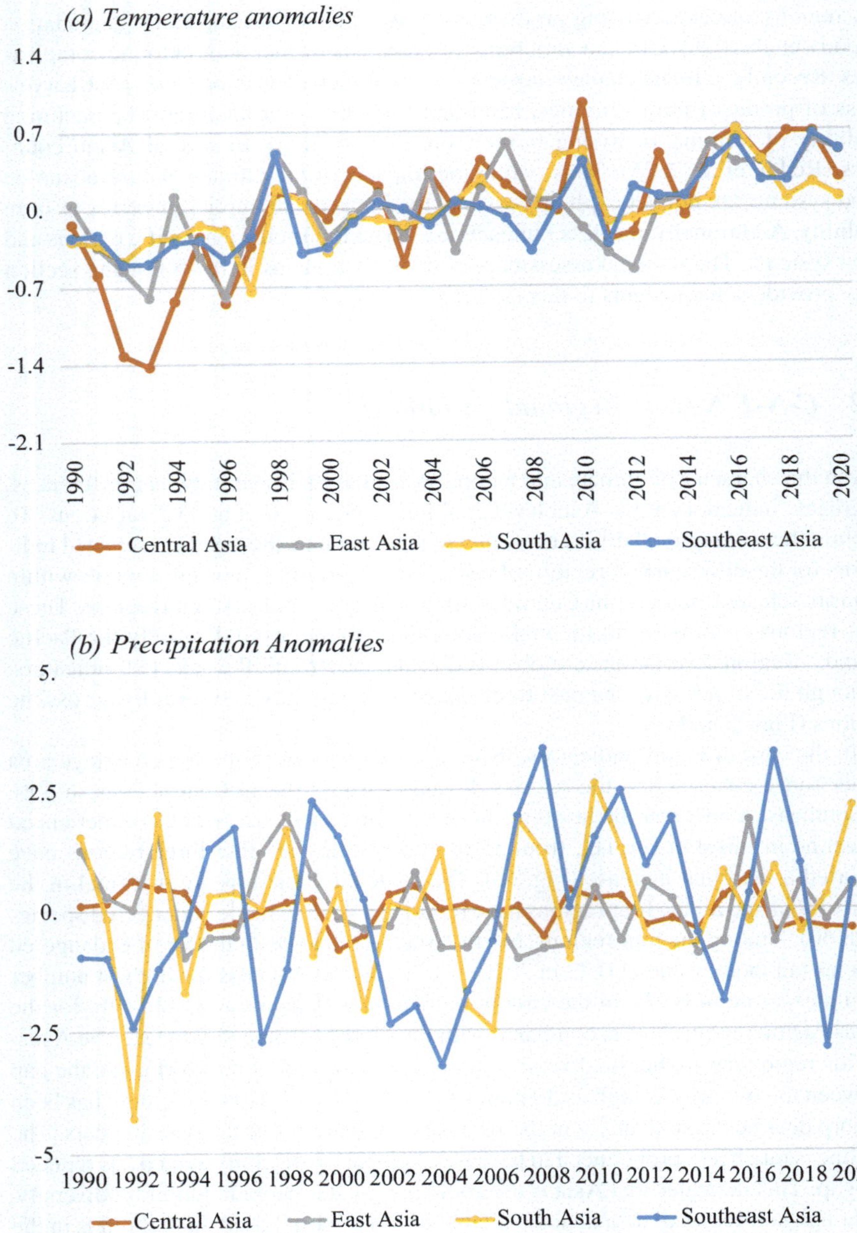

Fig. 5 Trajectories of climate indicators across Asian regions. *Source* https://ourworldindata.org/climate-change

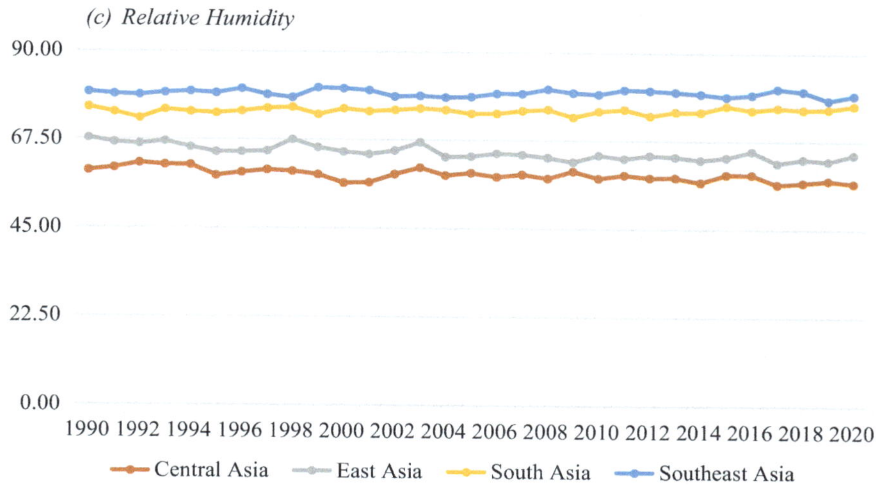

Fig. 5 (continued)

the top, while South Asia is at the bottom, in terms of all the indicators (Figs. 7a–d). The total per-capita daily calorie supply has increased in all regions, and the gap between them has narrowed over the period (Fig. 7a). The same is true for the extent of malnutrition, as shown by the relative position of South Asia between 2001 and 2020 (Fig. 7b). As far as the composition of the food basket is concerned, South Asia has the lowest supply/availability levels of animal protein, as well as the lowest share of animal protein in the total calorie supply. Despite the fact that the availability of animal protein has doubled in South Asia between 1990 and 2020, regional differences have increased over this period (Fig. 7c). The share of Animal protein in the total calories has remained low in all the regions, i.e., less than 10% (Fig. 7d). The share is as low as 3% or less. This has implications for emissions from food systems.

To examine the C-N-F nexus patterns at the sub-regional level, important C-N-F indicators are plotted separately for each region (Figs. 8, 9, 10 and 11). No clear and distinct pattern in the trends of C-N-F indicators could be observed in any of the sub-regions of Asia. In Southeast Asia, trendlines of forest cover and the share of animal protein appear to be moving in opposite directions (Fig. 11). East Asia recorded a lower increase in total calorie supply, with a close to stable forest cover (Fig. 9). This could be due to the fact that the impact of climate and nature is not reflected even at the sub-regional level.

While the trend lines indicate improvements and declines over the three-decade period, the year-on-year changes are not well captured, especially when there are fluctuations. Compound annual growth rates (CAGRs) are calculated for the important indicators of the C-N-F nexus using actual data on annual precipitation, percentage of forest cover, total calories supplied per capita per day, and proportion of agricultural land to the total geographical area. Sub-region-wise CAGR for different

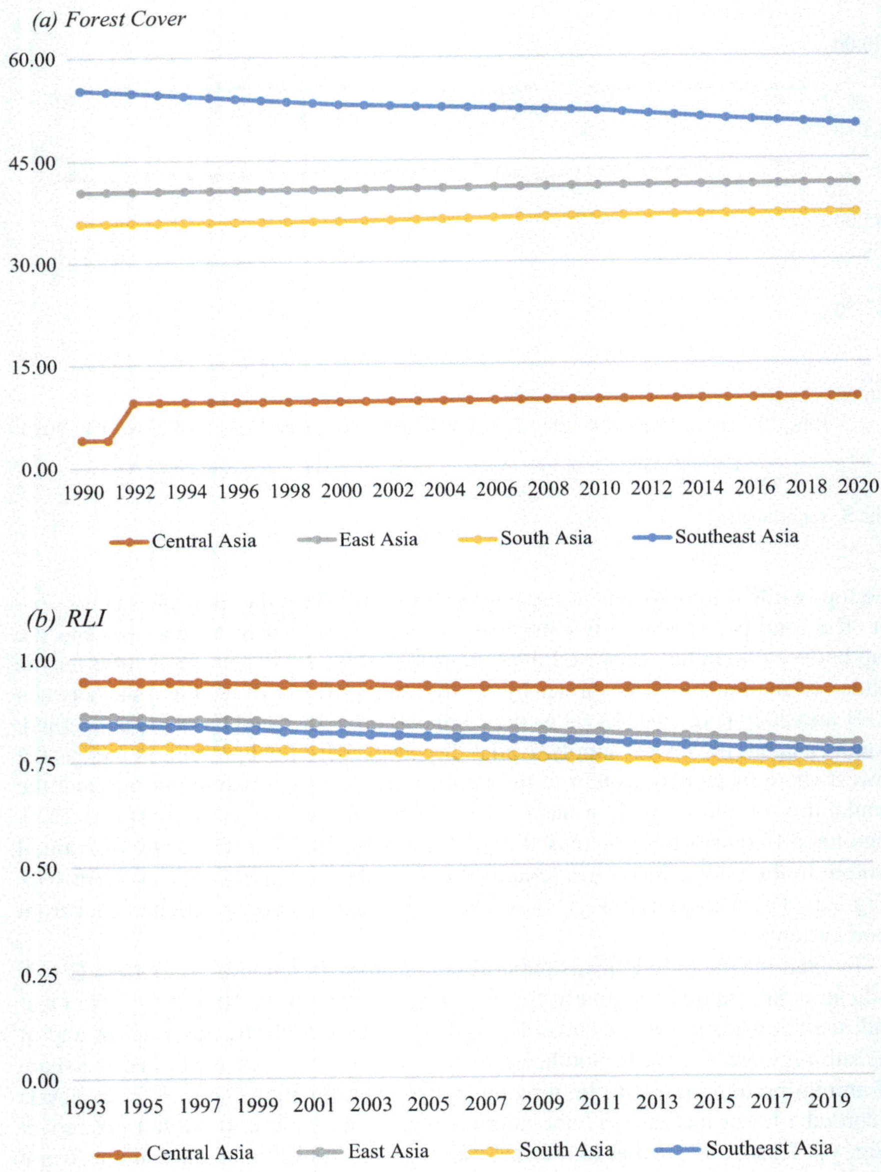

Fig. 6 Trajectories of nature indicators across Asian regions. *Source* https://ourworldindata.org/climate-change

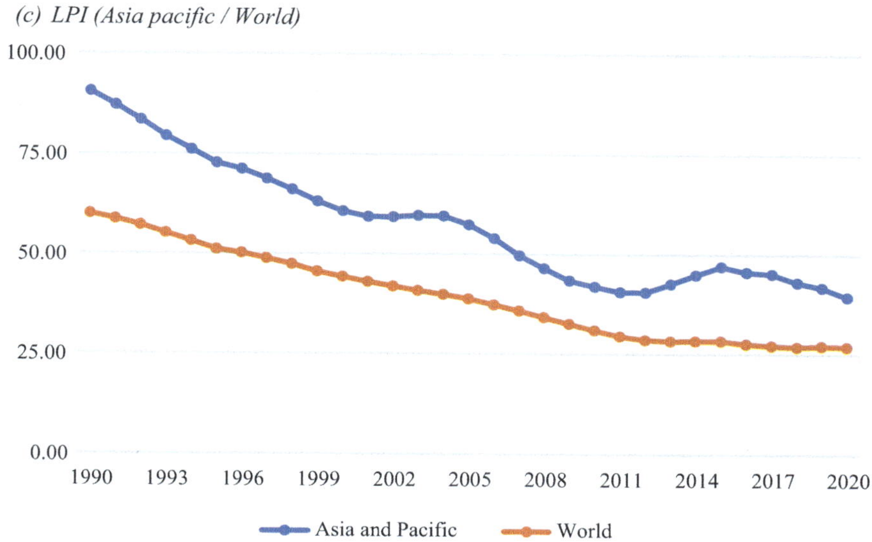

Fig. 6 (continued)

periods for these indicators is presented in Table 2. Although these CAGR figures do not provide any further evidence of a systematic pattern in the C-N-F nexus, they reveal some intrinsic details. While the average annual precipitation has shown positive growth over the 30-year period at the Asia level, this growth is limited to one sub-region (Southeast Asia) only, further highlighting the importance of contextualizing aggregate analysis with micro-assessments. The annual growth was positive in the first (1990–1999) and negative in the second (2000–2009) and third (2010–2019) decades. But the most recent decade (2010–19) has recorded high and negative growth in precipitation across all the sub-regions. These periodic growth rates indicate a declining future rainfall.

In the case of forest cover, Southeast Asia recorded negative growth, along with Asia, while Central, East, and South Asia have recorded positive growth. Only Central Asia has recorded high positive growth. Southeast Asia has reported negative growth in all the sub-periods. Of all the indicators, total calorie supply has reported consistently positive growth in all the sub-regions and sub-periods. The proportion of the area under agriculture, which determines the food supply (calorie supply), has reported high and positive growth in the Southeast and Central Asia sub-regions in most of the sub-periods. More than the growth in area under agriculture land, it is the GDP growth that might have helped keep pace with food (calorie) supplies. GDP growth has been consistently positive and high in all sub-regions, particularly over the last two decades (2000–2009 and 2010–2019).

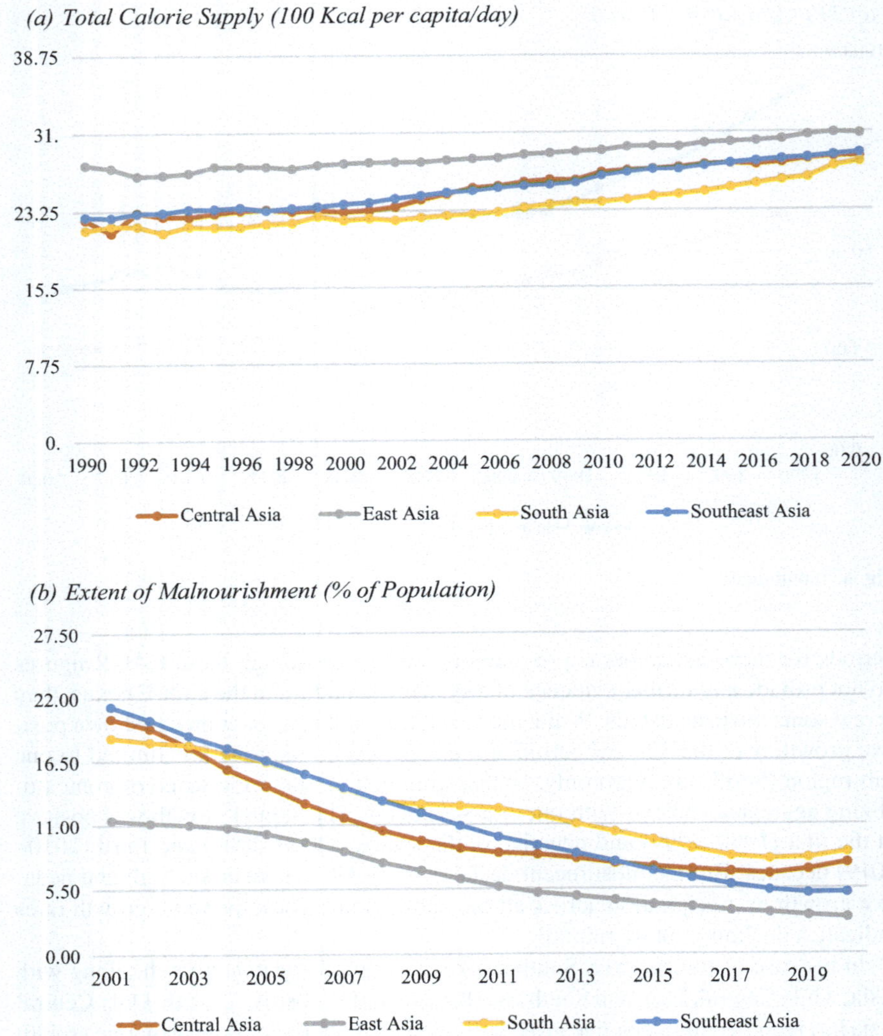

Fig. 7 Trajectories of food indicators across Asian regions. *Source* https://ourworldindata.org/climate-change

4 Conclusions

The preceding analysis provides a comprehensive overview of the C-N-F nexus, utilizing trend lines of various indicators that represent the three dimensions of C-N-F over a three-decade period at the Asia level and its four sub-regions. There are clear patterns that indicate climate change is evident in terms of temperature rise, and

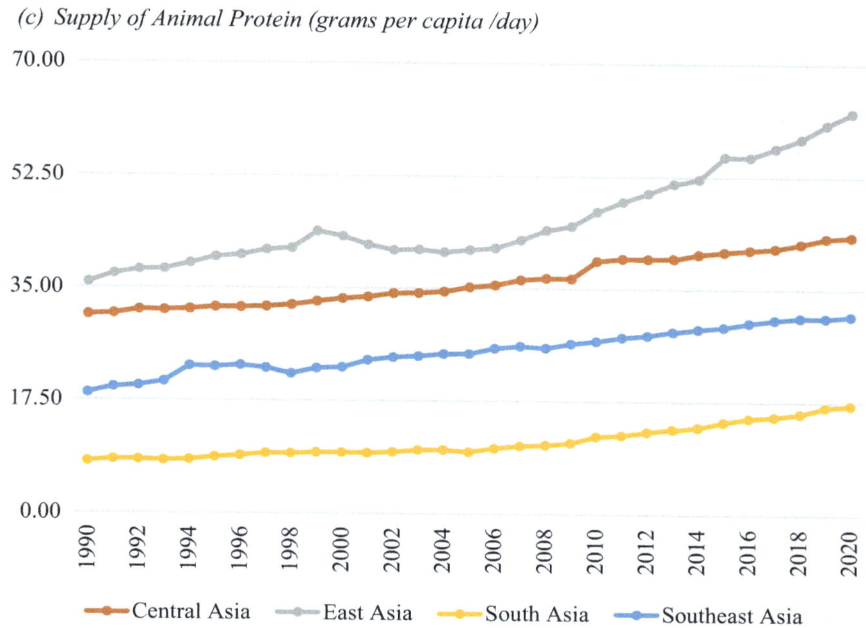

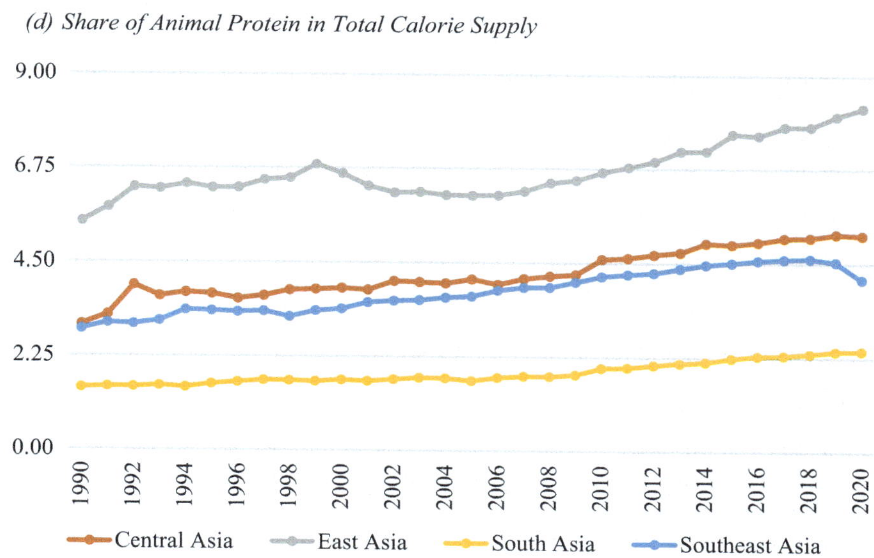

Fig. 7 (continued)

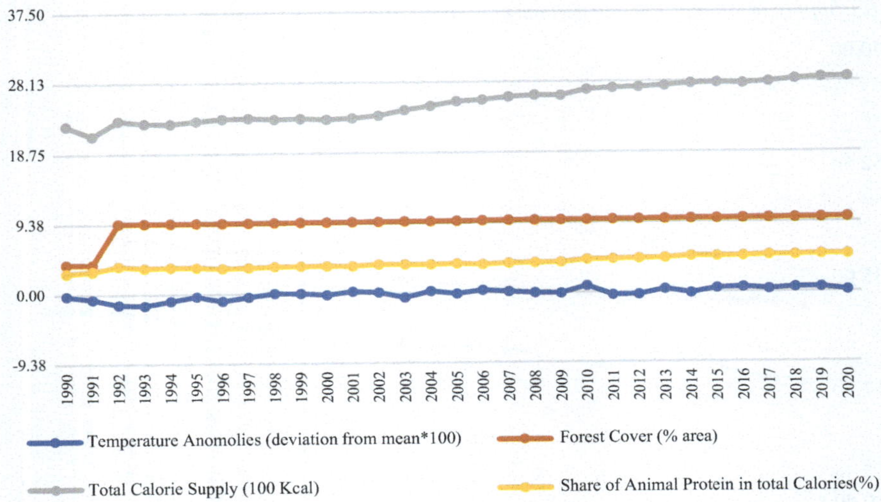

Fig. 8 Climate, nature, and food trajectories in Central Asia. *Source*: https://ourworldindata.org/climate-change

precipitation has, on average, marginally increased in Asia, but reported declining annual growth rates in three of the four regions. However, the most recent decade has recorded a negative annual growth in precipitation, which is indicative of future trends. The primary indicator of nature, biodiversity, has shown a substantial decline. Although the annual growth rate of forest cover declined between 1990 and 2020 in Asia and Southeast Asia, other regions have shown a marginal increase. The high growth rate in forest cover in Central Asia could be due to its very low base of forest cover. Although regional data on biodiversity is not available, the drastic decline in biodiversity in the Asia–Pacific region is expected to be reflected at the regional level. In the case of food indicators, total calorie availability per capita has increased over the years in all regions. The consumption of animal protein and its share in total calorie supply are on the rise across regions, although its share remains quite low in Asia and its sub-regions compared to developed economies—largely attributed to lower income levels rather than preferences.

The increasing food supply does not seem to be impacted by deteriorating climate and natural factors. While available research, based on laboratory studies or modeling, indicates clear linkages between climate and food, actual data do not support this hypothesis. This does not mean that there is no C-N-F nexus. Hitherto, research has been carried out in silos, i.e., assessing one or two systems at a time, such as climate or climate–food. In the process, the contribution of nature to the food system is bypassed or ignored. Food systems draw heavily from the natural systems in terms of soils, water, and other resources, which are supported by biodiversity and ecosystem services. Unless the degradation of the natural system reaches a flash point (soil quality, water stress, water quality, etc.), its impact on food production or

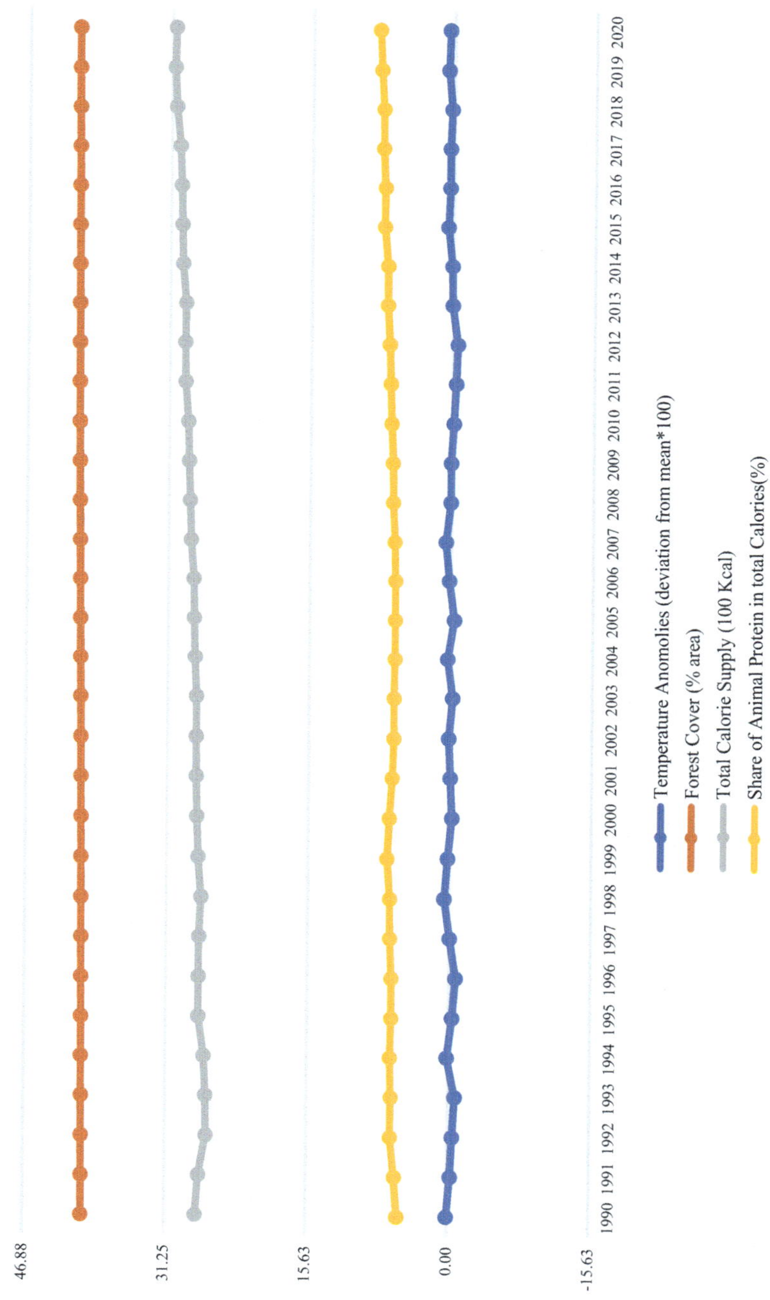

Fig. 9 Climate, nature, and food trajectories in East Asia. *Source* https://ourworldindata.org/climate-change

Fig. 10 Climate, nature, and food trajectories in South Asia. *Source* https://ourworldindata.org/climate-change

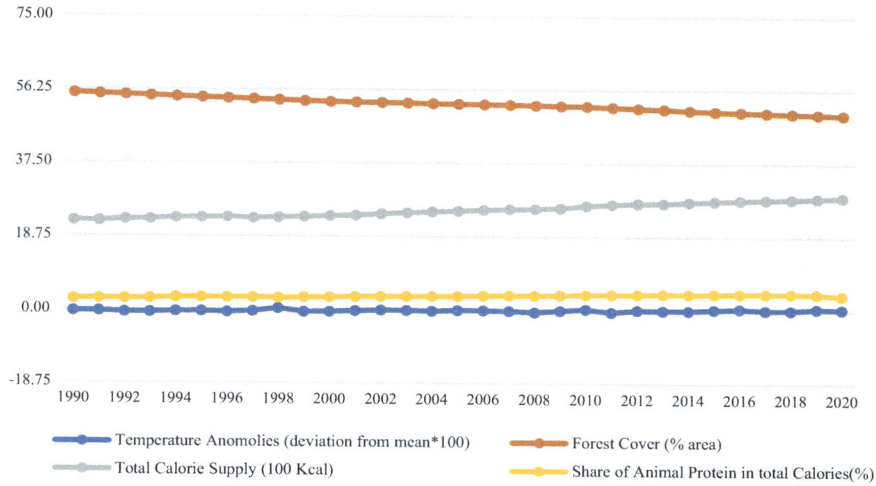

Fig. 11 Climate, nature, and food trajectories in Southeast Asia. *Source* https://ourworldindata.org/climate-change

supplies will not be conspicuous. More importantly, given the high policy priority of food security/self-sufficiency in all countries, technologies that increase land productivity are being promoted regularly. As a result, the excessive and imbalanced use of chemicals is keeping yields artificially stable or improving, as evidenced by the increasing levels of agricultural inputs over time (Reddy and Reddy 2007). Such practices are proving to be environmentally unsustainable and economically unviable. Although the impacts are yet to be visible at the aggregate level, they are evident in some countries and regions in Asia (Reddy and Reddy 2007; Reddy et al. 2025). This serves as the foundational evidence of the existence and the strong interactions between the facets of the C-N-F nexus.

To further flesh out the details of the C-N-F nexus, the assessment needs to be more comprehensive and requires a deeper analysis. A comprehensive analysis, in terms of the number of indicators and a deeper examination of their interlinkages, will reveal quantifiable relationships between the factors. Additionally, a more disaggregated assessment, i.e., at the country level, could provide important insights into the C-N-F nexus. The next part of the book will embark on a more detailed and rigorous quantitative analysis, i.e., identifying the number of relevant indicators and establishing their linkages across the sub-regions. Besides, the changes in the linkages over different periods will be examined.

Table 2 Compound annual growth rates of selected C-N-F nexus indicators

Period	1990–1999	2000–2009	2010–2019	1990–2020
Average annual precipitation (in cm)				
Asia	0.41	−0.27	−1.43	0.09
Central Asia	0.22	2.16	−1.04	−0.29
East Asia	−0.11	0.82	−0.80	−0.24
South Asia	−0.60	0.09	−0.92	−0.08
Southeast Asia	0.34	−0.27	−1.37	0.05
% of forest cover in total geographical area				
Asia	−2.50	−0.02	−0.14	−0.88
Central Asia	8.80	0.18	0.19	2.90
East Asia	0.06	0.10	0.05	0.08
South Asia	0.08	0.16	0.08	0.11
Southeast Asia	−0.35	−0.17	−0.37	−0.32
Daily calorie supply per capita (Kcal)				
Asia	0.30	0.90	0.73	0.71
Central Asia	0.32	1.25	0.55	0.79
East Asia	0.04	0.39	0.57	0.33
South Asia	0.63	0.75	1.37	0.88
Southeast Asia	0.43	0.84	0.98	0.78
% of agricultural land in total geographical area				
Asia	1.80	0.06	0.22	0.66
Central Asia	0.75	−0.02	0.14	0.25
East Asia	−0.02	−0.80	−0.24	−0.39
South Asia	−0.08	0.05	0.11	0.01
Southeast Asia	0.40	0.94	0.76	0.78
GDP/Capita (PPP)				
Asia	−0.11	2.04	2.12	1.36
Central Asia	−4.82	6.75	3.16	1.49
East Asia	1.97	2.28	1.90	2.21
South Asia	3.23	4.10	4.22	3.86
Southeast Asia	−0.66	0.23	1.18	0.31

Source Based on the data: https://ourworldindata.org/climate-change

References

ADB (2021) Asian development outlook update 2021—theme chapter: transforming agriculture in Asia. Asian Development Bank. https://www.adb.org/sites/default/files/publication/726556/ado2021-update-theme-chapter.pdf

Cheng L, Zhang X, Reis S et al (2022) A 12% switch from monogastric to ruminant livestock production can reduce emissions and boost crop production for 525 million people. Nat Food 3:1040–1051. https://doi.org/10.1038/s43016-022-00661-1

Chia SY, Lim MW (2022) IOP Conf Ser: Mater Sci Eng 1257:012001. https://doi.org/10.1088/1757-899X/1257/1/012001

Crippa M, Solazzo E, Guizzardi D et al (2021) Food systems are responsible for a third of global anthropogenic GHG emissions. Nature Food 2:198–209. https://doi.org/10.1038/s43016-021-00225-9

Dasgupta P (2021) The economics of biodiversity: the Dasgupta review. HM Treasury, London. https://assets.publishing.service.gov.uk/media/602e92b2e90e07660f807b47/The_Economics_of_Biodiversity_The_Dasgupta_Review_Full_Report.pdf

F4B (Finance for Bio-diversity) (2021) The climate-nature nexus: implications for the financial Sector. https://www.naturefinance.net/wp-content/uploads/2022/09/F4B_Climate-Nature-Nexus_Implications-for-Financial-Sector_20210527.pdf

FAO (2004) Socio-economic analysis and policy implications of the roles of agriculture in developing countries. http://www.fao.org/es/esa/roa/pdf/summary.pdf. Accessed on 16 Jun 2022

Fujimori S, Wu W, Doelman J et al (2022) Land-based climate change mitigation measures can affect agricultural markets and food security. Nature Food 3:110–121. https://doi.org/10.1038/s43016-022-00464-4

Hasegawa T, Sakurai G, Fujimori S et al (2021) Extreme climate events increase risk of global food insecurity and adaptation needs. Nature Food 2:587–595. https://doi.org/10.1038/s43016-021-00335-4

IPBES and IPCC (2021) IPBES-IPCC co-sponsored workshop report on biodiversity and climate change. https://doi.org/10.5281/zenodo.4782538

IPCC (2023) Climate change 2023: synthesis report. In: Core Writing Team, Lee H, Romero J (eds) Contribution of Working Groups I, II and III to the sixth assessment report of the intergovernmental panel on climate change. IPCC, Geneva, Switzerland, pp 35–115. https://doi.org/10.59327/IPCC/AR6-9789291691647

Mendelsohn R (2014) The impact of climate change on agriculture in Asia. J Integr Agri 13(4):660–665. https://doi.org/10.1016/S2095-3119(13)60701-7

McDonald AJ, Balwinder-Singh KA et al (2022) Time management governs climate resilience and productivity in the coupled rice–wheat cropping systems of eastern India. Nat Food 3:542–551. https://doi.org/10.1038/s43016-022-00549-0

OECD (2000) Multifunctionality: towards an analytical framework. [COM/AGR/APM/TD/WP(2000)3/REV2]. Accessed on 16 Jun 2022

Reddy VR (2003) Land degradation in India: extents, costs and determinants. Econ Polit Wkly XXXVIII(44)

Reddy VR (2005) Costs of resource depletion externalities: a study of groundwater over exploitation in Andhra Pradesh, India. Environ Dev Econ (Camb) 10(Part-IV)

Reddy VR, Behera B (2025) Tackling the current social, economic and ecological crises: in search of an alternative integrated system. Springer

Reddy VR, Rahut DB (2023) Multi-functionality of rice production systems in Asia: a synoptic review. Asian Development Bank Institute, Tokyo, Japan. https://www.adb.org/publications/multifunctionality-of-rice-production-systems-in-asia-a-synoptic-review

Reddy VR, Reddy PP (2007) Increasing costs in agriculture: agrarian crisis and rural labour in India. Indian J Labour Econ 50(2)

Reddy VR et al (2018) A water–energy–food nexus perspective on the challenge of eutrophication. Water 2018(10):101. https://doi.org/10.3390/w10020101

Reddy VR, Rahut DB, Sonobe T (2025) Future of small holder farms in Asia. Asian Development Bank Institute, Tokyo, Japan. https://www.adb.org/sites/default/files/publication/1041036/future-smallholder-farming-asia.pdf

Sawan ZM (2018) Climatic variables: evaporation, sunshine, relative humidity, soil and air temperature and its adverse effects on cotton production. Inf Process Agric 5(1):134–148. https://doi.org/10.1016/j.inpa.2017.09.006

Sun Z, Scherer L, Tukker A et al (2022) Dietary change in high-income nations alone can lead to substantial double climate dividend. Nat Food 3:29–37. https://doi.org/10.1038/s43016-021-00431-5

Xia L, Cao L, Yang Y et al (2023) Integrated biochar solutions can achieve carbon-neutral staple crop production. Nat Food 4:236–246. https://doi.org/10.1038/s43016-023-00694-0

Xu X, Sharma P, Shu S et al (2021) Global greenhouse gas emissions from animal-based foods are twice those of plant-based foods. Nat Food 2:724–732. https://doi.org/10.1038/s43016-021-00358-x

Yan X, Qing-shi M, Jie G, Xiang-fang T, Hong-fu Z (2017) Effects of relative humidity on animal health and welfare. J Integr Agric 16(8):1653–1658. https://doi.org/10.1016/S2095-3119(16)61532-0

Zhu J, Luo Z, Sun T et al (2023) Cradle-to-grave emissions from food loss and waste represent half of total greenhouse gas emissions from food systems. Nat Food 4:247–256. https://doi.org/10.1038/s43016-023-00710-3

Part II
Regional, Sub-regional, and Country Perspectives

Chapter 3
C-N-F Nexus in the Asian Context: A Quantitative Assessment

1 Introduction

Theoretical literature and studies based on scientific research and modeling logically indicate that a C-N-F nexus is a given phenomenon, due to the strong linkages between the three dimensions. Although understanding the theoretical basis of the nexus is crucial for comprehending the linkages, it is not sufficient to empirically establish these linkages. While the preceding chapter, with the help of visualization, demonstrates the existence of a core nexus, it is limited to the trajectories and co-movement of the base indicators of the nexus. The visualizations, while helpful in illustrating them, do not establish the significance of the linkages. Under what biophysical, socioeconomic, and policy environments does the nexus intensify and transform over time and space? How imbalances in the nexus, arising from the factors mentioned above, manifest in real-world outcomes. Hence, it is necessary and valuable to establish this relationship quantitatively, especially for drawing policy attention and designing targeted policies. This part of the book provides a quantitative assessment of the C-N-F nexus with the help of a more detailed analysis incorporating a larger number of variables. While this chapter presents a detailed assessment of the nexus in the context of Asia and its four sub-regions, a country-wise assessment of selected countries representing each of the four sub-regions of Asia is taken up in the remaining chapters of this part. This chapter is organized into four sections. The following section presents the approach and the variables identified for analysis, along with the statistical methods used to quantify the linkages. The section further on (three) provides the analysis for Asia and its four sub-regions. Section 4 presents concluding remarks and provides the rationale for identifying and selecting the case study countries that represent the four sub-regions.

V. R. Reddy et al., *Strengthening the Climate–Nature–Food Nexus*,
https://doi.org/10.1007/978-3-032-29423-4_3

2 Data and Methods

Several variables representing all three vertices of the nexus have been identified (Table 1). Time series data for a period of 31 years, from 1990 to 2020, inclusive of both years, have been collected. A total of 20 indicators have been identified for analysis, and the details of the selected indicators, along with their important descriptive statistics, are presented in Table 1. The data is gathered from open-access public data sources like "Our World in Data," World Bank, etc. In the case of "climate" data, information is available for three core indicators: temperature, precipitation, and humidity. All three indicators are included in the present analysis. In the case of precipitation, both absolute measures (average annual precipitation) and annual precipitation anomalies are included, while in the case of temperature, data on anomalies are available, rather than absolute temperatures (average annual temperature). The measurement of "nature" indicators is particularly challenging, given their broad definition and cross-cutting impacts, such as water, soil, biodiversity, and land use. That is, in order to satisfactorily represent the various facets of nature (degradation or restoration), a comparatively large number of variables for this factor had to be considered. Five different indicators related to nature, including forest cover, water availability, water stress, irrigation water, and land use, are included. In the case of "food," availability, composition, and quality indicators are considered, viz., the supply of total calories per capita, protein supply, the consumption/share of animal protein in the total calorie supply, and the percentage of malnourished population, which are included to represent "food."

In addition to the C-N-F-specific variables, economic variables such as GDP are included to understand the sensitivity of the nexus indicators to economic development across the sub-regions over a span of three decades. Variables representing farm intensification, such as water use for agriculture, fertilizer, and pesticide use, are also included to assess the impact of agriculture technology and farm intensification on the food system. All indicators are standardized using percentages, population, and land as appropriate. For instance, all the "nature" indicators are expressed in percentage terms, such as forest cover and water availability. "Food" indicators are standardized using population, i.e., per-capita availability, and so is GDP. And intensification indicators are standardized with land, i.e., per unit of land.

Correlation coefficients are estimated for all these indicators to establish the linkages between each of the indicators and the others. Correlation coefficients help in understanding the direction and strength of these linkages. These coefficients indicate the nature and strength of the relationships among climate, nature, and food as well as how the indicators representing the C-N-F influence each dimension. Theoretically, each indicator is expected to influence climate, nature, and food in a particular way in the Asian context. For instance, the climate indicator average annual rainfall is likely to have a positive influence on all three dimensions "C-N-F." A higher average rainfall is expected to improve the climatic conditions through improved natural habitats. It also enhances food availability by improving land productivity through increased moisture availability and enhanced nutrient absorption. At the same time, improved

Table 1 Variables considered along with their descriptive statistics for Asia

Variable name	Abbreviation	C-N-F	Descriptive statistics	
			Mean	SD
Annual average precipitation in mm	AARF in mm	Climate	1498	69
Average annual precipitation anomaly—yearly deviation from the mean of 1990–2020	AARF anomaly in mm	Climate	0.6	72
Average annual temperature anomaly in centigrade—deviation from the mean of 1990–2020	AATP anomaly in C	Climate	0.0	0.4
Relative humidity in %	RH in %	Climate	69	0.6
Annual freshwater withdrawals for agriculture use	%FWWD-AGRL	Nature	78	03
Level of water stress—freshwater withdrawal as a proportion of available freshwater resources	WSTR in %	Nature	42	03
Agricultural land as % of total area	% AL	Nature	41	02
Area under forest as % of total area	FC in %	Nature	33	02
Renewable internal freshwater resources per capita (cubic meters)	FW/CAP in M^3	Nature	12,289	2350
Daily total calorie supply per person (kcal)	TCS/day/person in kcal	Food	2553	197
Daily total protein supply per person (grams)	TPS/capita/day in grams	Food	72	09
Animal Protein available per capita per day in grams	APS/capita/day in grams	Food	27	06
Share of the animal protein in the total daily calorie supply in %	SAP in %	Food	4.1	0.5
Percentage of malnourished population in total population (%)	% MP	Food	14	05
Per-capita GDP in 2015 USD in purchasing power parity terms	GDP/capita/year in US$	Economic	5363	943

(continued)

Table 1 (continued)

Variable name	Abbreviation	C-N-F	Descriptive statistics	
			Mean	SD
Nutrient nitrogen use in kilograms per hectare of crop land	Nitrogen use in kgs/ha	Intensification	61	09
Nutrient phosphate use in kilograms per hectare of crop land	Phosphate use in kgs/ha	Intensification	25	03
Nutrient potash use in kilograms per hectare of crop land	Potash use in kgs/ha	Intensification	21	2
Total fertilizer consumption (kilograms per hectare of cropland)—sum of nitrogen, phosphate, and potash	NPK use in kgs/ha	Intensification	106	11
Pesticide use in kilograms per hectare	Pesticide use in kgs/ha	Intensification	03	0.2

Source Authors

climatic conditions and biodiversity (nature) together would have a positive influence on crop/food production. On the other hand, the influence of rainfall and temperature anomalies on the C-N-F dimensions can be either positive or negative, depending on their increasing/declining trend relative to the averages. Increased rainfall over the base year could have a positive influence on all three dimensions and a negative influence in the case of a decline. The opposite holds in the case of temperature anomaly, i.e., an increased temperature over the average level is expected to have a negative influence in the case of an increase and a positive influence in the case of a decline on all three aspects. Relative humidity is expected to have a positive or negative influence, depending on its current level (see Chap. 2).

Of the nature indicators, water withdrawal for agriculture and water stress are expected to have a negative association with climate and nature and a positive influence on food. At the same time, water stress beyond a certain point is expected to have a negative impact on food. Whereas, per-capita freshwater availability would have a positive influence on all three dimensions. The share of agriculture land would have a negative association with climate and nature. Negative linkage due to unsustainable farming practices, coupled with land-use changes from forest to agricultural land, is theoretically valid. For the same reason of land-use changes, it has negative linkages with temperature anomalies, as agriculture land often increases at the cost of forest cover. And it also has a positive influence on food production. Forest cover, a primary indicator of nature and biodiversity, has a positive impact on climate and the environment. However, its relationship is dependent on the indicator of climate. For instance, forest cover will have a positive impact on rainfall anomalies, while it will have a negative impact on temperature anomalies. Similarly, it has a positive relation with

Table 2 Theoretically expected signs of the selected indicators

Indicator	Theoretically expected signs of correlation coefficients		
	Climate	Nature	Food
AARF in mm	+	+	+
AARF anomaly in mm	+/−	+/−	+/−
AATP anomaly in C	+/−+	+/−	+/−
RH in %	+/−	+/−	+/−
% FWWD-AGRL	−	−	+
WSTR in %	+	−	+/−
FW/CAP in M^3	−	+	+
% AL	+	−	+
FC in %	+/−	+/−	+/−
TCS/day/person in kcal	+/−	+/−	+
TPS/capita/day in grams	+/−	+/−	+
APS/capita/day in grams	−	−	+
SAP in %	−	−	+
% MP	−	−	−
GDP/capita/year in US$	−	−	+
Nitrogen use in kgs/ha	−	−	+
Phosphate use in kgs/ha	−	−	+
Potash use in kgs/ha	−	−	+
NPK use in kgs/ha	−	−	+
Pesticide use in kgs/ha	−	−	+

Source Based on the studies reviewed in this volume (Chaps. 1 and 2)

freshwater availability and a negative association with water stress. Typically, forest cover has a positive influence on food systems through its ecosystem services and biodiversity benefits. Given the trade-off between forest cover and agricultural land, forest cover could have a negative influence on food production, as the expansion of agricultural land could be constrained by forest cover.

In the case of food indicators, the total calorie supply or availability influences food systems positively, while it may have a positive or negative influence on climate and nature, depending on the adoption of sustainable agriculture practices. That is, unsustainable intensive farm practices would increase production but degrade the natural resources and systems that adversely impact the climate. While total protein supply indicates the composition and quality of food supply, its influence on climate and nature depends on its source. In the case of plant-based protein, it would have a positive impact on the climate and nature. On the other hand, the supply of animal-based protein (APS/SAP) has a negative impact on the climate and nature. Widespread malnutrition is expected to have a negative influence on all three dimensions, as degraded nature and adverse climate could reduce food production and increase malnutrition.

While economic indicators, such as GDP, along with intensification variables (the use of fertilizers and pesticides), have a positive influence on food systems, they have a negative impact on climate and nature. GDP growth often puts pressure on the environment through consumerism, which directly increases GHG emissions and results in greater climate risks. Similarly, use of fertilizers and pesticides as part of farm intensification degrades the natural system (soil, water, and biodiversity loss) and aggravates climate risks while contributing positively to production and food availability.

Correlation matrices have been constructed using all the indicators. While the complete correlation matrices are presented in the appendix, this section only presents and discusses the significant coefficients. The correlation matrices for all these variables are estimated across four time periods using the time series data for 31 years, i.e., period 1 (1990–2020); period 2 (1990–2000); period 3 (2001–2010), and period 4 (2011–2020), and for Asia as a whole along with its four sub-regions. Before proceeding with the analysis, however, it is essential to note some caveats regarding the data and analysis.

Data availability—certain variables, such as water stress, are not available for the entire time series under consideration; this has understandably impacted the 31-year estimates of the correlation matrices. Time sensitivity of variables—the long time frame considered has both merits and challenges. The 31-year time frame is significant in terms of the interactions between climate, nature, and food. This time period is marked by a global (and particularly in the Asian context) boom in the food supply and security levels. It is also marked by unprecedented technological advancements that have enabled the aggressive expansion of food availability across the world and the sub-regions considered in this study. However, this also meant that the effects of these changes—technological and policy—have been significant within this time period as well. Within each decade, the impact of shifting policy and technology has meant that the correlations between these variables have shifted considerably.

Subjectivity and geographic specificity—while the broad picture painted by the data appears consistent across the regions, the statistical analysis also highlights the region-specific conditions and associated impacts. In some cases, these may seem counterintuitive or inconsistent with the data in other sub-regions. However, when contextualized within the specific conditions of the sub-region, it is clear that the nature of interactions among C-N-F factors is consistent. This is also due to the indirect impacts of various indicators that tend to vary among sub-regions. These challenges are addressed within the study by considering the overall Asia aggregates, in tandem with sub-regional correlations, and primarily focusing on the 31-year aggregate correlations, along with the last decade (2011–2020) to account for the recency of the data.

3 C-N-F Nexus: Asia Aggregate Analysis

Correlation coefficients are estimated for the Asia region as a whole for different periods. The estimates of correlation coefficients at the Asia aggregate level revealed that several selected indicators proved to be significant. The findings not only vindicate the notion that the overall relationship across the C-N-F nexus can be measured quantitatively, but also that the interactions align with our earlier qualitative assessment (Chap. 2). That is, not only is there concrete quantitative evidence showing the linkages between climate, nature, and food, but the correlation matrices also enable us to tease out the interactions between the vertices. Besides, the strong interrelations between them across highly varied geographic, socioeconomic, demographic, and climatic conditions are also evident. The analysis also helps identify significant variables for constructing a composite C-N-F index, which aims to measure the strength of interactions between the C-N-F vertices, and provides the basis for monitoring the nexus across regions in the future.

The strength of these interactions is not entirely cancelled out at the aggregate level, further supporting the rationale and purpose of this study. The correlation coefficients are consistent with the theoretical and qualitative expectations of these interactions. Although some counterintuitive linkages are observed, they bring out interesting insights into understanding the direct and indirect nature and multiplicity of interactions among the C-N-F factors, i.e., the externalities associated with the impacts. Linkages between the core indicators of C-N-F, viz., average annual temperature anomaly (AATP anomaly), forest cover (FC in %), and total calorie supply/capita/day in kcal (TCS/day/person in kcal), and different indicators are identified for different periods. Only the significant correlation coefficients are presented and discussed here (Table 3).

3.1 Climate

The long-term (1990–2020) correlation coefficients indicate that temperature anomalies are linked negatively with freshwater availability, relative humidity, forest cover, and the extent of malnutrition, and positively with GDP and total calorie supply (Table 3). All these relationships are in the expected lines. That is, GDP, representing economic development, and temperature anomalies are positively linked, i.e., both move in the same direction. Similarly, total calorie supply moves in the same direction as temperature anomalies due to the former's (TCS) strong positive association with GDP. This is the reason why the deteriorating climate (increased temperature) and the extent of malnutrition move in opposite directions. Forest cover and freshwater availability tend to move in opposite directions with temperature anomalies. In the case of relative humidity, it can be in either direction, depending on the current levels. It is worth noting that indicators with significant associations differ from period to period. In period 2 (1990–2000), only NPK use, pesticide use,

Table 3 Significant correlation coefficients of the C-N-F nexus indicators in the Asia region

Period	Climate—temperature anomaly (AATP anomaly)	Nature—forest cover (FC in %)	Food—total calorie supply (TCS/day/person in kcal)
P1 (1990–2020)	FW/CAP in M^3: −0.64*; FC in %: −0.29***; GDP/capita/year in US$: 0.73*; RH in %: −0.63*; TCS/day/person in kcal: 0.75*; %MP: −0.71*	FW/CAP in M^3: 0.83*; GDP/capita/year in US$: −0.33**; SAP in %: −0.53*;; WSTR in %: 0.31***; TCS/day/person in kcal: −0.42**; %MP: 0.36***; AATP Anomaly in C: −0.29***; RH in %: 0.42**	AATP anomaly in C: 0.75*; RH in %: −0.61*; FC in %: −0.42**; FW/CAP in M^3: −0.83*; NPK: 0.75*; WSTR: −0.89*; % AL: 0.54*; %MP: −0.99*
P2 (1990–2000)	NPK use in kgs/ha: 0.67*; Pesticide use in kgs/ha: 0.44***; TCS/day/person in kcal: 0.52**	GDP: 0.52**; SAP in %: −0.92*; % AL: −0.99*; FW/CAP in M^3: 0.96*; TCS/day/person in kcal: −0.56**	AATP anomaly in C: 0.52**; RH in %: −0.70*; FC in %: −0.56**; FW/CAP in M^3: −0.75*; NPK use in kgs/ha: 0.85*; % AL: 0.63**
P3 (2001–2010)	FW/CAP in M^3: −0.44***; FC in %: −0.48***; NPK use in kgs/ha: 0.59**; WSTR in %: −0.45***; GDP/capita/year in US$: 0.45***; RH in %: −0.52**; TCS/day/person in kcal: 0.50**; %MP: −0.46	GDP/capita/year in US$: −0.98*; SAP in %: −0.90*; FW/CAP in M^3: 0.99*; WSTR in %: 0.90*; TCS/day/person in kcal: −0.99*; %MP: 1.00*; AARF in mm: −0.69*; AARF anomaly in mm: −0.69*; AATP anomaly in C: −0.48***	AATP Anomaly in C: 0.50**; AARF Anomaly in mm: 0.69*; FC in %: −0.99*; FW/CAP in M^3: −0.99*; AARF anomaly: 0.69*; %MP: −0.99*
P4 (2011–2020)	FW/CAP: −0.80*; FC: −0.82*; NPK: 0.76*; WSTR: 0.54**; % AL: 0.69*; GDP/capita/year in US$: 0.81*; TCS/day/person in kcal: 0.74*; %MP: −0.87*	NPK use in kgs/ha: −0.95*; GDP/capita/year in US$: −0.98*; SAP in %: −0.96*; %AL: −0.92*; FW/CAP in M^3: 1.00*; TCS/day/person in kcal: −0.99*; %MP: 0.97*;AATP anomaly in C: −0.82*; RH in %: 0.48***	AATP Anomaly in C: 0.74*; RH in %: −0.52**; FC in %: −0.99*; FW/CAP in M^3: −0.99*; NPK use in kgs/ha: 0.92*; pesticide use in kgs/ha: 0.69*; % AL: 0.96*; %MP: −0.93*

Note *, **, and *** indicate the level of significance at 1%, 5%, and 10% levels, respectively
Source Estimated by the authors

and total calorie supply turned out to be significant, with a positive association with temperature anomaly. This period is associated with increased production and use of chemicals in crop production, resulting in higher yields across Asian countries. Temperature anomalies, which are on the rise, are associated with the increased use of chemicals in farming.

During the period 2001–2010 (period 3), a larger number of indicators turned out to be significant. When compared to the earlier periods (P1 and P2), water stress

(WSTR in %) turned out significant with a negative sign. That is, temperature anomalies and water stress tend to move in opposite directions, although they are expected to move in the same direction. This could be due to the development of water supply infrastructure, including groundwater, during this period. Interestingly, the relationship between these two turned positive in period 4 (2011–2020), indicating that the development of water supply infrastructure, especially groundwater, could not keep pace with demand. That is, temperature anomalies and water stress have been moving in the same direction in recent years. During period 4, the percentage of agricultural land to total land (% AL) has turned out to be significant with a positive sign. The trade-off between forest cover and agriculture land is the reason. Besides, farm intensification practices (like the use of NPK and pesticides) are positively associated with temperature anomalies. It is worth noting that all the indicators, except water stress, exhibit consistent trends over the periods. Besides, the strength of the coefficients has increased over the periods.

3.2 *Nature*

In the case of forest cover (nature), freshwater availability per capita, water stress, extent of malnutrition, and relative humidity are positively associated in period 1 (1990–2020). While water availability and relative humidity are expected to move in the same direction with forest cover, their positive association with water stress and the extent of malnutrition requires some explanation. These positive linkages could be due to indirect impacts, such as the trade-off between forest cover and agricultural land in the case of malnutrition. In the case of water stress, the explanation is again the development of water supply infrastructure that often outpaces the replenishing mechanisms, such as forest cover. Forest cover (nature) and temperature anomalies (climate) are moving in opposite directions, which again establishes the nexus between climate and nature. GDP and forest cover move in opposite directions, as economic development is realized at the cost of natural capital, especially in developing countries. The negative association between forest cover and total calorie supply (food) is due to the trade-off between forest and agricultural land use. This is reflected in the strong negative association between forest cover and the share of agricultural land during period 2 (1990–2000). As observed in other studies (see Chap. 2), the consumption of animal protein is expected to degrade nature many times over when compared to plant-based protein. All these indicators appear in all the later periods with the same signs and improved strength. Additionally, the use of NPK turned out to be significant, with a negative sign. This indicates that agricultural intensification through chemical farming and forest cover loss moves in opposite directions, further illustrating the nexus. The only aberration is the negative association between forest cover and rainfall, as well as rainfall anomaly, which is limited to a single period (2001–2010).

3.3 Food: Cost of Feeding Asia

Food systems, as measured in terms of total calorie supply, are positively associated with temperature anomalies, NPK use, and agricultural land. Malnutrition, water stress, water availability, relative humidity, and forest cover are moving in the opposite direction of total calorie supply. The positive linkage between NPK use and total calorie supply is expected and obvious. It is interesting to note that temperature anomaly and food supply are moving in the same direction. This could be due to the fact that temperature anomalies are reaching a level that can impact crop production. This is reflected in the negative association with water stress and availability, indicating that the temperature rise is being offset by an increased water supply, resulting in decreased water stress and increased water availability. And the negative association between forest cover and relative humidity is in the expected direction. In period 3 (2001–2010), rainfall and rainfall anomaly have shown a positive relation with calories. As in the case of climate and nature, the strength of these relationships has increased in the recent decade (2011–2020) compared to earlier decades. Given the importance of food systems among the three dimensions, the association between total calorie supply, on the one hand, and the extent of malnutrition and other indicators needs further understanding.

The correlation coefficient between calorie availability and the percentage of the malnourished population is highly negative. While this is on the expected lines, there is considerable discourse on the value of measuring calorie availability in isolation from nutritional requirements. Large populations may have diets that are calorically sufficient, but nutritionally imbalanced or deficient. This could be due to an imbalanced food basket (excessive carbs), excessive chemical use, poor processing, unhygienic eating conditions, etc. The data herein suggest that while it may be true, it may not be particularly the case within the Asian context, given the highly negative correlation coefficients (above −0.90) in all periods[1] (Table 4). Furthermore, it is noteworthy that the average temperature anomaly, which represents the annual temperature deviation from the mean temperature (1990–2020), is also highly correlated with both calorie availability (positively, at 0.75) and malnutrition (negatively, at −0.71). A similar pattern is seen across all the sub-periods as well, suggesting that positive temperature anomalies (marginally higher temperatures) are currently aiding in calorie availability (Mendelsohn 2014). However, the data also highlights the importance of the delicate balance within climate indicators. That is relative humidity, which is positively linked to higher average annual precipitation. In contrast, precipitation anomaly (particularly in the recent decade) is negatively related to temperature anomalies. This implies a cooling effect and, therefore, has a negative correlation with calorie availability. Besides, the increased rainfall, while leading to better access to water (higher freshwater withdrawals for agriculture at 0.49 and freshwater availability per capita at 0.48), might not always lead to better caloric availability. This

[1] The non-significance of the correlation coefficient during 1990–2000 (P1) is due to non-availability of data on the extent of malnutrition.

could be due to the changed distribution and intensity of rainfall, resulting in crop losses, including post-harvest losses.

The data at the aggregate level tells the story of continued intensification. It is clear, particularly during the 2011–2020 period, that food availability comes at a great cost to nature, as evident from the highly negative correlation between caloric supply and the percentage of forest area. It, however, also shows that technological advancement, as indicated by the high negative correlation with freshwater withdrawals for agriculture (−0.84), and the high positive correlation with fertilizer and pesticide use. The increased caloric availability (and lower malnutrition rates) over the years is largely driven by the expansion in the share of agricultural land, which evidently comes at the cost of a decrease in forest area (−0.99). This is further compounded by the evidence that the share of agricultural land is highly negatively correlated with the share of forest area (−0.92 in the last decade and −0.98 for the overall period).

It is essential to note that this is not limited to agricultural produce. Over time, with increasing incomes (GDP), dietary composition shifts to include higher levels of protein, in particular, animal protein. The animal protein supply chain is far more detrimental to ecological balance (Cheng et al., 2022; Xu et al., 2021). It is particularly interesting that the increasing availability of animal protein is also characterized by an increased agricultural land share (Table 5). This supports the thesis that, despite changing dietary composition, it is not a one-for-one replacement of calories

Table 4 Correlation coefficients of food and other indicators (Asia aggregate 2011–2020)

Indicator	Calorie supply/day/person in kcal	% of malnourished population
TCS/day/person in kcal	1.00	
% MP	−0.93*	1.00
AARF in mm	−0.30	0.30
AARF anomaly in mm	−0.28	0.29
AATP anomaly in C	0.74*	−0.87*
RH in %	−0.52**	0.44***
FC in %	−0.99*	0.97*
% FWWD-AGRL	−0.84*	0.70*
FW/CAP in M^3	−0.99*	0.97*
Nitrogen use in kgs/ha	0.92*	−0.85*
Phosphate use in kgs/ha	0.67*	−0.78*
Potash use in kgs/ha	0.59**	−0.71*
NPK use in kgs/ha	0.92*	−0.91*
WSTR in %	0.15	−0.31
Pesticide use in kgs/ha	0.69*	−0.42***
% AL	0.96*	−0.82*

Note *, **, and *** indicate the level of significance at 1%, 5%, and 10% levels, respectively
Source Estimated by the authors

Table 5 Correlation coefficient of total protein supply/cap/day in grams (2011–2020)

Indicator	Correlation coefficient
TCS/day/person in kcal	0.99*
% MP	−0.97*
AARF anomaly in mm	0.80*
FC in %	−0.99*
%FWWD-AGRL	−0.79*
FW/CAP in M^3	−0.99*
Nitrogen use in kgs/ha	0.93*
Phosphate use in kgs/ha	0.74*
Potash use in kgs/ha	0.66*
NPK use in kgs/ha	0.95*
Pesticide use in kgs/ha	0.60**
% AL	0.92*

Note *, **, and *** indicate the level of significance at 1%, 5%, and 10% levels, respectively
Source Estimated by the authors

from agrarian products with animal protein. It is rather a supplement on top of the better availability of plant-based sources of nutrition and calories. This may also be attributed to the rising GDP per capita, which results in higher affordability and increased availability of animal sources of protein in these economies.

4 Conclusions

The interlinkages of various indicators establish a clear relationship between expanded food availability and nature. Furthermore, they also establish the acceleration of climatic variability/risks, as well as the degradation of natural capital and biodiversity. Specifically, forest cover has a strong negative correlation with temperature anomalies (−0.82). Intensification also correlates with higher levels of degradation, with increased nitrogen usage correlating to higher levels of forest cover loss due to changing land-use patterns.

The preceding analysis also corroborates the evidence of climate–nature interactions. Freshwater availability per capita is positively correlated with the percentage of forest cover (0.83 across the 30-year timeframe), with a stronger correlation established in the recent period (2011–2020) of 0.99. Infrastructural progress is also recognizable in the correlation values, with both freshwater withdrawals for agriculture and freshwater availability per capita being highly correlated across both the 30-year and 2011–2020 periods, with a much stronger correlation in the 2011–2020 period (0.68 versus 0.81).

Closing the loop refers to the stress that climatic variance places on food production systems, contributing to higher water stress and increased water withdrawals (not to mention potential crop losses from untimely rains or flooding). Higher levels of forest cover correlate with higher freshwater withdrawals for agriculture (0.81 in 30-year period and 0.78 in the last decade, 2011–2020), indicating a greater need for freshwater extraction to meet food demands. It is also curious that high levels of freshwater withdrawals per capita are negatively correlated with calorie availability. This suggests that higher freshwater withdrawal and availability may also coincide with other outcomes that lead to lower calorie availability. Restoration of natural capital (such as expanding or preserving forest areas) may alleviate climate variability. The analysis clearly indicates that there is still potential for agriculture in the region to benefit from temperature increases (Mendelssohn 2014), as evidenced by the correlations between calorie availability and temperature anomalies, relative humidity, average annual precipitation, and precipitation anomalies. While early gains of farm technology were limited, the last decade (2011–2020) saw the accumulation of gains from continued intensification (in some areas) and the resulting returns from the scale application of intensification technologies and infrastructure.

Thus, within the Asia aggregates, there is clear evidence of the climate–nature–food nexus. These forces, in some cases, alleviate the negative impacts of certain variables, but largely work together to magnify the negative impacts of exploitative activities or technologies. It is especially important to note that some of the more restorative interactions between two of the factors may have detrimental impacts on the third factor in the nexus—as in the case of restorative C-N interactions leading to lower F values, higher C-F values leading to lower N values, and high F-N values resulting in lower C values. This now leaves us with the need to understand how the C-N-F nexus is being handled in different regions and specific countries, given the varying prevailing circumstances, and the resulting outcomes in terms of the C-N-F factors. These aggregate (Asia) findings need to be checked for their congruence at the sub-region and country levels. The country analysis will delve into the specifics of C-N-F nexus interactions within the sub-regions and countries, focusing on outcomes and key takeaways for managing this matrix to ensure positive outcomes.

4.1 Identifying Case Study Countries

In what follows, we discuss the basis for identifying the individual countries that represent each sub-region. For this, two variables, namely, natural resource availability (forest cover) and socioeconomic development (GDP), are considered. These two indicators, across three levels—low, medium, and high—result in a matrix into which each of the sub-regions can be classified (Table 6). While each combination of the matrix has its own value in terms of analysis, sub-region and country-level analysis can be restricted to those with particularly high vulnerability or relevance in terms of economic activity. It can be postulated that these two may be closely related, as economic activity usually contributes to nature degradation, increasing vulnerability to imbalances in the C-N-F nexus.

Based on the two indicators, the four sub-regions are categorized as follows: low-low (Central Asia), high-high (East Asia), medium-medium (South Asia), and low–high (Southeast Asia). Using the same criteria, four countries representing each sub-region are identified. These include: Uzbekistan from Central Asia; Japan from East Asia; Nepal from South Asia; and Vietnam from Southeast Asia (Table 7). While Uzbekistan falls into the low-low category, Japan is in the high-high category, Nepal is in the medium–low category, and Vietnam is in the high-low category. It is worth noting that we have deliberately avoided selecting the most influential big countries of China and India, as they are geographically vast with varying agro-climatic conditions. This makes the analysis less representative of the country as a whole. Additionally, these countries have been extensively researched and are therefore less likely to provide new insights.

Table 6 Three-way matrix of resource availability and socioeconomic development

		Socioeconomic development		
		Low	Medium	High
Baseline resource availability	Low	Central Asia (Uzbekistan)		
	Medium		South Asia (Nepal)	
	High	Southeast Asia (Vietnam)		East Asia (Japan)

Table 7 Selected countries representing the four sub-regions and three-way matrix

Baseline natural resources	Socioeconomic development		
	Low	Medium	High
Low	Uzbekistan		
Medium	Nepal	Vietnam	
High			Japan
	Region/Country	Forest cover	Per-capita GDP
	Asia	32.25	6903
	Uzbekistan	8.4	3249
	Japan	68.4	34,650
	Korea	64.5	31,378
	India	24.3	1806
	Bhutan	71.4	3092
	Nepal	41.6	1031
	Laos	71.9	2529
	Vietnam	47.2	3303
	China	23.4	10,358

References

Cheng L, Zhang X, Reis S et al (2022) A 12% switch from monogastric to ruminant livestock production can reduce emissions and boost crop production for 525 million people. Nat Food 3:1040–1051. https://doi.org/10.1038/s43016-022-00661-1

Mendelsohn R (2014) The impact of climate change on agriculture in Asia. J Integr 13(4):660–665. https://doi.org/10.1016/S2095-3119(13)60701-7

Xu X, Sharma P, Shu S et al (2021) Global greenhouse gas emissions from animal-based foods are twice those of plant-based foods. Nat Food 2:724–732. https://doi.org/10.1038/s43016-021-00358-x

Chapter 4
Central Asia Case Study: Uzbekistan

1 Introduction

The Central Asia sub-region falls in the 'low-low' category of natural capital and socioeconomic development (GDP). Central Asia comprises ten main countries: Afghanistan, Armenia, Azerbaijan, Georgia, Kazakhstan, Kyrgyzstan, Pakistan, Tajikistan, Turkmenistan, and Uzbekistan. Most of these countries that gained independence from the former Soviet Union are in transition. These countries face climate risks and vulnerabilities associated with their geographical characteristics, such as deserts, semi-deserts, and monoculture farm practices. This region is heavily dependent on agriculture, with approximately 50% of the population relying on it. More importantly, Central Asia faces climate risks, including increasing temperatures and decreasing rainfall, across the entire region, especially in the western parts of Turkmenistan, Uzbekistan, and Kazakhstan (Lioubimtseva and Henebry 2009). Even the recent data for the region vindicates these trends (Chap. 2). Given the low natural resource base and increasing climate risks, farming is increasingly becoming difficult in certain areas of the region. This is resulting in a slowdown of the economies and an increase in food insecurity. In fact, Central Asia is experiencing large-scale out-migration, which is termed environmental migration. Resource degradation-triggered migration is more specific to the central Asian region within Asia. This makes assessing and understanding the C-N-F nexus in this region critical for addressing the socioeconomic development problems of the region (Artem et al. 2020).

2 Quantitative Assessment of C-N-F Nexus: Central Asia

It is worth noting, at the outset, that Turkmenistan is excluded from the analysis due to data constraints. Central Asia is a region with low rainfall (<600 mm) and low forest cover (9.5%) (Table 1). Due to the low forest cover, 55% of the land is used

V. R. Reddy et al., *Strengthening the Climate–Nature–Food Nexus*,
https://doi.org/10.1007/978-3-032-29423-4_4

Table 1 Descriptive statistics of selected indicators for Central Asia

Indicator	C-N-F	Descriptive statistics	
		Mean	SD
AARF in mm	Climate	598.7	38
AATP Anomaly in C	Climate	0	0.6
RH in %	Climate	58.8	1.3
% FWWD-AGRL	Nature	84.1	4.2
WSTR in %	Nature	71.8	1.2
% AL	Nature	55.3	1
FC in %	Nature	9.5	1.4
FW/CAP in M^3	Nature	4154	570
TCS/day/person in kcal	Food	2531	222
TPS/capita/day in grams	Food	76.2	9.1
APS/capita/day in grams	Food	28.2	6.3
SAP in %	Food	4.2	0.6
% MP	Food	14.2	5.7
GDP/capita/year in US$	Economic	2460	912
Nitrogen use in kgs/ha	Intensification	48.8	20
Phosphate use in kgs/ha	Intensification	10	3.2
Potash use in kgs/ha	Intensification	2.7	1.3
NPK use in kgs/ha	Intensification	61.5	21.3
Pesticide use in kgs/ha	Intensification	0.8	0.3

for agriculture. It is a highly water-stressed region (71.8%). Agriculture accounts for 84% of the total water use. Farm intensification appears to be low, with minimal fertilizer application and negligible use of pesticides. This is reflected in the extent of malnourishment (14%) and poverty levels, which are around 30% (Artem et al. 2020). These indicators suggest that Central Asia is among the most vulnerable regions in terms of C-N-F dimensions. The number of indicators representing the three dimensions has turned out to be significant in all four periods (Table 2). The signs of the coefficients not only support the C-N-F nexus observed in the trend visualization (Chap. 2) and the quantitative assessment at the aggregate (Asia) level. That is, aggregation across sub-regions has not significantly affected the direction and strength of the C-N-F nexus linkages.

2.1 *Climate*

As expected, rainfall (both absolute and anomaly) and relative humidity move in the opposite direction to temperature anomalies (Table 2). While food indicators—total

Table 2 Significant correlation coefficients of the C-N-F nexus indicators for the Central Asia sub-region

Period	Climate—temperature	Nature—% forest cover	Food—total calorie supply/capita/day (kcal)
P1 (1990–2020)	RH in %: −0.69*; NPK use in kgs/ha: 0.45**; WSTR in %: 0.49**; Pesticide use in kgs/ha: 0.40**; GDP/capita/year in US$: 0.58**;% MP: –0.62*; TCS/day/person in kcal: 0.64*; AARF in mm: –0.28***; AARF anomaly in mm: –0.28***; SAP in %: 0.58**	FW/CAP in M^3: 0.86*; %SAP: –0.58**; TCS/day/person in kcal: 0.51**;%MP: –0.38***; RH in %: –0.30***;% FWW-agrl: –0.89*; NPK use in kgs/ha: 0.41**; pesticide use in kgs/ha: 0.70*;% AL: 0.86*	AATP anomaly in C: 0.65*; RH in %: –0.55**; FC in %: 0.51**; % FWWD-AGRL: –0.68*; NPK use in kgs/ha: 0.83*; WSTR in %: 0.73*; pesticide use in kgs/ha: 0.83*; GDP/capita/year in US$: 0.91*; SAP in %: 0.95*; MP in %: –0.98*
P2 (1990–2000)	NPK use in kgs/ha: 0.60**; RH in %: –0.71*; AARF: –0.50**; AARF anomaly in mm: –0.50**	FW/CAP:1.00*; NPK use in kgs/ha: 0.68*; GDP/capita/year in US$: –0.93*; % SAP: –0.93*;TCS: 0.83*; % FWWD-AGRL: –0.94*; pesticide use in kgs/ha: 0.99*; % AL: 0.98*	FC: 0.83*; FW/CAP in M^3: 0.83*; NPK: 0.52*; % AL: 0.86*; pesticide use in kgs/ha: 0.78*; GDP: –0.72*; SAP in %: 0.69*; % FWWD-AGRL: –0.78*
P3 (2001–2010)	% FWWD-AGRL: 0.41***; RH in %: −0.69*; NPK use in kgs/ha: 0.48***; % AL: –0.58**; SAP in %: 0.44***	FW/CAP in M^3: −1.00*; NPK: 0.78*; Pesticide use in kgs/ha: 0.77*; % AL: –0.41***; GDP/capita/year in US$: 0.99*; SAP in %: 0.82*; TCS: 0.98*; % FWWD-AGRL: –0.98*; %AL: –0.41***; MP in %: –0.99*	FC in %: 0.98*; FW/CAP in M^3: –0.98*; NPK: 0.72*; pesticide use in kgs/ha: 0.86*; GDP/capita/year in US$: 0.97*; SAP in %: 0.78*; % FWWD-AGRL: –0.96*; MP in %: –0.99*
P4 (2011–2020)	FW/CAP in M^3: –0.68*; FC in %. 0.66*; NPK use in kgs/ha: 0.70*; WSTR in %: 0.62**; pesticide use in kgs/ha: –0.40***; GDP: 0.74*; TCS/day/person in kcal: 0.62**;%MP: –0.83*; % FWWD-AGRL: 0.42***; SAP in %: 0.67*	FW/CAP: −1.00*; NPK: 0.93*; GDP: 0.97*; SAP in %: 0.96*; %AL: 0.69*; TCS: 0.98*;%MP: –0.77*; AATP anomaly in C: 0.66*; RH in %: –0.55**	AATP anomaly in C: 0.62**; RH in %: –0.56**; FC in %: 0.98*; FW/CAP in M^3: −0.98*; NPK use in kgs/ha: 0.89*; % AL: 0.72*; %MP: –0.74*; GDP/capita/year in US$: 0.97*; SAP in %: 0.95*; % FWWD-AGRL: –0.89*

Note *, **, and *** indicate the level of significance at 1, 5, and 10% levels, respectively
Source Estimated by the authors

calorie supply and share of animal protein—are positively associated with temperature anomalies, the extent of malnutrition is negatively associated. This could be due to changes in the microenvironment (such as water, sanitation, and hygiene) associated with temperature anomalies. The strong linkages between the microenvironment and malnutrition are well established in the literature (IRC et al. 2013). These linkages are much stronger in central Asia due to its exposure to air and water pollution (ADB 2012). Water stress (a natural indicator), GDP (an economic indicator), along with NPK and pesticide use (intensification), have revealed a positive association with temperature anomalies. The indicators and their signs are consistent over the periods. However, in period 4 (2011–2020), more indicators turned out to be significant, with freshwater availability per capita becoming significant with a negative sign. That is, fresh water availability is likely to be negatively associated with any increase in temperature, especially in low-rainfall regions like Central Asia.

2.2 Nature

Central Asia is characterized by very low forest cover (about 10 percent), which is a main indicator of its natural state. This could be the reason why forest cover and percent of agricultural land (%AL) are moving in the same direction (positive sign). Water availability, NPK use, pesticide use, and total calorie supply have a positive association with forest cover (Table 1). The positive association between these indicators is as expected. Whereas the share of agriculture water use (% FWWD-AGRL), the share of animal protein, the extent of malnutrition, and relative humidity have a negative relationship with forest cover. The strength of these relationships has increased over the years. In period four (2011–2020), the temperature anomaly proved to be significant, with a positive sign indicating that forest cover and temperature anomalies are moving in the same direction, particularly during the recent decade. This may be due to the policy support for afforestation in the region, which has resulted in an increase in forest cover from 4% during the early 1990s to about 10% by the late 2010s.

2.3 Food

Given the climate and nature vulnerabilities in the region, food security remains a major challenge. Total calorie supplies are positively associated with temperature anomalies, forest cover, input use (NPK and pesticides), water stress, GDP, and the share of animal protein (SAP, expressed as a percentage). The positive association between temperature anomalies and total calorie supply is similar to that of Asia. The same holds true for the negative relationship between total calorie supply and the extent of malnutrition. In contrast, the association between GDP and total calorie supply is positive and consistent in all the periods except one (period 2—1990–2000). Another observation that needs explanation is the negative relation between total calorie supply and the share of water withdrawal for agriculture. These counterintuitive relations are discussed in the following section.

2.4 *Extreme Environments and the C-N-F Nexus*

For Central Asia, all the factors within the C-N-F nexus are under higher stress than for the rest of Asia. As illustrated by the descriptive statistics, the region has much higher water stress, significantly lower natural capital, and food availability is on the higher side. The source of calories is far more closely correlated with the consumption of animal protein, indicating a problem with agricultural output and, consequently, a nutritional imbalance.

While the correlation coefficients between temperature and food are relatively weaker in Central Asia compared to the Asian aggregate (0.65 vs. 0.75), they are much stronger in the case of relative humidity (0.82 vs. 0.09) (Table 3). These are largely attributed to the climatic conditions prevalent in the region. It is evident from the nations comprising this sub-region that the climatic conditions are predominantly arid and face high climate risks, including nuclear-related ones. The data reveals this geo-climatic information through anomalous correlations of certain factors when compared to the Asian aggregates. This is most evident in the correlations between forest cover, calorie availability, and the extent of malnutrition. The correlation coefficients are counter to the Asian aggregate values (Table 4).

It may not be immediately clear at the outset the gains in terms of relative humidity and rainfall anomalies from improved forest areas (Table 5). They directly translate to better calorie availability, as opposed to the lower availability due to detrimental outcomes. One important reason is that the region is dependent on food imports. Besides, this could be a result of an overall greening/cooling effect, making the region more conducive to agricultural activities as a whole. To this end, it is not irrational to understand that a positive feedback loop exists between agricultural activity and improved green cover, especially in the context of low forest cover. Besides, a climatic

Table 3 Correlation coefficients between different indicators of climate and food in Asia and Central Asia

Period	AATP anomaly in C to TCA		R H in % to AARF in mm	
	Asia	Central Asia	Asia	Central Asia
1990–2020	0.75*	0.65*	0.09	0.82*
2011–2020	0.74*	0.62*	0.59**	0.87*

Note *, **, and *** indicate the level of significance at 1, 5, and 10% levels, respectively
Source Estimated by the authors

Table 4 Correlation coefficients between nature (% FC) and food variables in Asia and Central Asia

Period	Asia		Central Asia	
	TCA/day/person in kcal	MP in %	TCA/day/person in kcal	MP in %
1990–2020	–0.42***	0.36**	0.51**	–0.38***
2011–2020	–0.99*	0.97*	0.98*	–0.77*

Note *, **, and *** indicate the level of significance at 1, 5, and 10% levels, respectively
Source Estimated by the authors

shift in temperatures and precipitation is trending towards making conditions more suitable for higher levels of agricultural output.

As is the case with the Asia aggregates, caloric availability is greatly driven by the intensification of agriculture and bolstered by an increased availability of animal protein (Table 6). This is especially the case given the lower baseline of soil quality and resource availability. Hence, higher levels of inputs (compared to other, more moist, and fertile sub-regions) are required to achieve comparable outputs (Reddy et al. 2025). These anomalous findings are particularly evident in the water stress and intensification indicators, further establishing the interactions of lower baseline values that contribute to counterintuitive correlations for the same indicators.

The water stress and share of agriculture land indicators for Central Asia are particularly interesting. The Asia aggregate water stress indicator has shown a highly negative correlation coefficient, while Central Asia has a highly positive correlation coefficient (Table 6). The share of agricultural land in Central Asia has no significance ($p > 0.10$), whereas the Asia aggregate has a positive and significant correlation coefficient. This suggests that, despite the importance of agriculture in this region, the primary sources of calories remain non-agrarian and trade-based. In contrast to the rest of Asia, where high water stress contributes to lower food output, and hence lower calorie availability. Most agriculture in Central Asia is conducted under conditions of high water stress, resulting in a positive correlation with calorie supply (Table 6). High water stress is the main reason for low farm intensification (low NPK use) despite the strong positive linkages between NPK, especially nitrogen, and total calorie supply (Table 6). Finally, the correlation coefficients for freshwater withdrawals

Table 5 Correlation coefficients of nature (% FC) and climate variables in Asia and Central Asia

Indicator/Region/Period	Asia		Central Asia	
	1990–2020	2011–2020	1990–2020	2011–2020
	FC in %	FC in %	FC in %	FC in %
AARF in mm	–0.29***	–0.82*	–0.12	–0.31***
RH in %	0.42**	0.48***	–0.30***	–0.55**
AATP anomaly in C	–0.29***	–0.82*	0.23***	0.66*

Note *, **, and *** indicate the level of significance at 1, 5, and 10% levels, respectively
Source Estimated by the authors

Table 6 Correlation coefficients of food, nature (% FC), and intensification variables in Asia and Central Asia (1990–2020)

Indicator	TCS/day/person in kcal	
	Asia	Central Asia
Nitrogen use in kgs/ha	0.93*	0.89*
Phosphate use in kgs/ha	–0.09	–0.08
Potash use in kgs/ha	0.08	0.01
NPK use in kgs/ha	0.75*	0.83*
WSTR in %	–0.89*	0.73*
Pesticide use in kgs/ha	–0.14	0.83*
Share of agriculture land (%)	0.54**	0.19

Note *, **, and *** indicate the level of significance at 1, 5, and 10% levels, respectively
Source Estimated by the authors

for agriculture and freshwater per capita to calorie availability complete the picture (Table 7). In climates such as those in Central Asia, water scarcity necessitates making difficult choices between providing water to the populace and utilizing it for agricultural purposes. When subsistence levels of access to water are threatened, agriculture takes a back seat to survivability. In Asia, the data show a high positive correlation between freshwater withdrawals for agriculture and freshwater per capita, indicating that infrastructure for water withdrawals benefits both agrarian activity and direct consumption. It also points to the fact that if water is available for direct consumption, there is usually enough for the limited agricultural activities. Given the low baselines of water availability in Central Asia, the data show a strong inverse relationship (Table 7).

The rest of the indicators align closely with the values for Asia. The analysis at the sub-regional level shows that the relationship between climate, nature, and food is consistent with the aggregate (Asia) picture. While this is certainly true, there are a few counter-indicative relationships that need to be understood in order to ensure the specific needs and circumstances of sub-regions are not lost. Central Asia is the strongest outlier in terms of the correlations between variables. This is understandable, given that it is the furthest outlier in terms of climatic conditions, biodiversity,

Table 7 Correlation coefficients between food (TCS) and nature (water) indicators in Asia and Central Asia

1990–2020	TCS/day/person in kcal	
Indicator	Asia	Central Asia
% FWWD-AGRL	–0.24***	–0.68*
FW/CAP in M^3	–0.83*	0.03
Indicator	% FWWD-AGRL	
FW/CAP in M^3	0.68*	–0.62**

Note *, **, and *** indicate the level of significance at 1, 5, and 10% levels, respectively
Source Estimated by the authors

and economic indicators, subject to much higher levels of stress across all these indicators, and with inconsistent policymaking and execution within the region due to political instability. It also poses the greatest challenge in terms of drawing meaningful comparisons with the rest of the sub-regions in Asia. Nevertheless, it is our firm belief that there is much to learn from understanding the circumstances of socioeconomic development in this sub-region, and simply considering it "too far of an outlier" threatens important takeaways. That is, understanding how to manage extreme climatic and natural resource stress—an eventuality that the rest of Asia might have to face one day. The needs of Central Asia are unique, and the outcomes are uncertain, as it continues to be one of the most vulnerable regions to climate change. Hence, the lessons learned from safeguarding the C-N-F nexus would be most helpful in addressing the problems and benefiting the people of Central Asia. In this regard, further narrowing the assessment to the country level may provide more insights. The assessment of the C-N-F nexus in the context of Uzbekistan is taken up as a case study in the next section.

3 C-N-F Nexus: Case of Uzbekistan

While Central Asia is at the extreme when compared to other sub-regions of Asia in terms of resource endowments, climate risks, and socioeconomic development, Uzbekistan presents the worst-case scenario among the Central Asian countries. Resource use allocation, such as land use and land cover, as well as water resources development, is often driven by the complex interplay of the above aspects. It was observed that the expansion of farming (food production), coupled with industrial progression (economic development), has placed immense pressure on the natural system. This has led to adverse impacts on biodiversity (forest cover) and ecosystem services (Karimov et al. 2023). The policy-driven (socioeconomic priorities) nature of the food nexus has resulted in soil and water resource degradation, exacerbating climate risks. Poor economic performance has constrained Uzbekistan from investing in the rejuvenation of natural systems, such as the Aral Sea, afforestation, and sustainable farm practices, as well as infrastructure and technology upgrades (efficiency and abatement). Poor, inefficient, and deteriorating irrigation and drainage infrastructure is impacting the quality and quantity of drinking water due to reduced river flows and pollution of groundwater sources. While the poor are more vulnerable to environmental risks, continued land degradation due to salinization and waterlogging has an adverse impact on rural livelihoods. In fact, it has been observed that unemployment at the regional level is correlated with air and water pollution (ADB 2012).

When compared to Central Asia, Uzbekistan is characterized by extreme natural and climatic conditions (Table 8). In terms of natural capital, it has only 7% forest cover, which is among the lowest in Asia. It receives the lowest rainfall of 248 mm. Average annual rainfall anomalies over the past 30 years are marginal, with high year-on-year variations, and average annual temperature anomalies are stable. However, one needs to examine the trends in climate indicators to assess the severity of the problem. The low rainfall is reflected in the availability of water and water stress. Freshwater availability is substantially lower, i.e., about 1/6 of Central Asia. And water stress in Uzbekistan is twice that of Central Asia. In fact, water stress at 150% indicates that present water extractions have reached unsustainable levels. Such high water stress is the reason for the degradation of surface and groundwater resources. Dependence on agriculture is more pronounced in Uzbekistan, with over 60% of the land allocated to agriculture. Agriculture accounts for more than 90% of the available water. Despite high water stress, agricultural intensification in terms of NPK use is more than double that of Central Asia. This explains the relatively better availability of total calories, as well as a lesser extent of malnutrition, in Uzbekistan compared to Central Asia. But its per capita GDP is substantially lower in Uzbekistan, indicating relatively lower overall economic development. Even the current levels of economic development are unlikely to continue in the face of deteriorating natural capital due to excessive pressure on resources. In 2021, Uzbekistan embarked on a massive afforestation program, planting 1 billion trees and shrubs with the objective of increasing forest cover to 30%.

Table 8 Descriptive statistics of the selected C-N-F indicators for Uzbekistan

Indicator	C-N-F	Descriptive statistics	
		Mean	SD
AARF in mm	Climate	248	44
AARF anomaly in mm	Climate	0.3	44.2
AATP anomaly in C	Climate	0	0.7
RH in %	Climate	52.1	2.3
% FWWD-AGRL	Nature	91.7	1.2
WSTR in %	Nature	149.9	6.6
% AL	Nature	61.2	2.9
FC in %	Nature	7.2	0.8
FW/CAP in M^3	Nature	621.8	84
TCS/day/person in kcal	Food	2740	339
SAP in %	Food	4.4	0.6
% MP	Food	11.1	7.3
GDP/capita/year in US$	Economic	1895	719
Nitrogen use in kgs/ha	Intensification	122.9	36
Phosphate use in kgs/ha	Intensification	36.7	10.7
Potash use in kgs/ha	Intensification	11.2	6.8
NPK use in kgs/ha	Intensification	170.8	42.4

3.1 C-N-F Nexus: Trends

In the context of extreme conditions in Uzbekistan, precipitation anomalies exhibit higher year-on-year fluctuations compared to temperature anomalies, while relative humidity remains more stable (Fig. 1a, b). Temperature anomalies are showing an increasing trend, while rainfall anomalies are on a declining trend. While the temperature anomaly has remained within a narrow range over the past 30 years, the rainfall anomaly has recorded a decline from +10 mm in 1990 to –30 mm in 2020. Rainfall also exhibited high fluctuations between years, ranging from −65.6 mm to +104 mm over a 30-year period. Relative humidity is also showing a declining trend (Fig. 1b). Altogether, the climate indicators are showing deteriorating climatic conditions in Uzbekistan. It is worth noting that temperature anomalies have remained in the positive range since 1996. The temperature trends are commensurate with the IPCC predictions on global warming. But the aggregate temperature anomalies remain below 1 °C.

Five indicators, namely, forest cover, agricultural land use, water availability, water use for agriculture, and water stress, are considered to represent nature. All the indicators, except forest cover, have shown a declining trend over the past 30 years (Fig. 2). Forest cover has increased marginally, from 6 to 8%, over the same three-decade period. This increase in forest cover has come at the cost of a decline in the share of agriculture land. The decline in the share of agricultural land is significantly greater than the increase in forest cover. This could be due to land-use and land-cover changes, such as for residential and industrial purposes. Policy changes aimed at protecting forests and increasing forest cover appears to have some impact, though not a significant one. On the other hand, the decline in agriculture land could be due to the increased water constraints. Per-capita water availability has declined substantially, impacting the share of water use for agriculture. Although water stress has declined from 153 to 143% over the period, stress levels remain high.

Trajectories of total calorie supply, the share of animal protein in total protein, and the extent of malnutrition, representing the food dimension of the nexus, are used to assess hunger/poverty, nutrition, and the composition of food. Over the past 30 years, all indicators have shown improvement, which is in line with Asia and its sub-regions (Fig. 3). The total calorie supply has increased from 2711 kcal to approximately 3246 kcal per capita/day over the 30 year period. The share of animal protein in the total protein supply has gone up from 3.8 to 5.7% over the period. The improved calorie supply, coupled with the increase in the share of animal protein, has resulted in a substantial decline in the extent of malnutrition, from 18 percent in 1990 to 2.5% of the total population in 2020. It is worth noting that such a sharp decline occurred between 2001 and 2020, as malnutrition data are available only from 2001 onwards. The improved food status, despite severe natural resource constraints and climate risks, is quite commendable. This could be attributed to the proactive food policies, including farm intensification, a shift to non-plant-based calories, and trade policies.

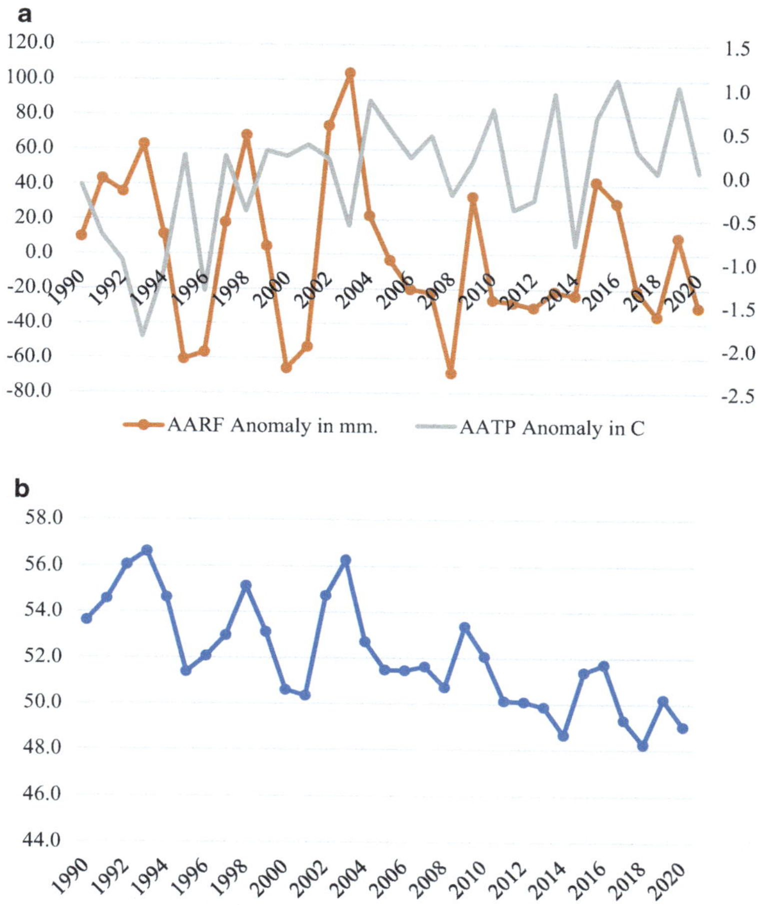

Fig. 1 **a** Trajectories of rainfall and temperature anomalies (climate) in Uzbekistan. *Source* https://ourworldindata.org/climate-change. **b** Trajectories of relative humidity (climate) in Uzbekistan. *Source* https://ourworldindata.org/climate-change

To assess the C-N-F nexus trajectories, the core indicators of the three dimensions are presented visually (Fig. 4). The trajectories of the three dimensions over the past three decades in Uzbekistan clearly show that the climate is characterized by increasing temperatures and declining precipitation. At the same time, nature, represented by forest cover, shows an improvement; however, the extent of forest cover in Uzbekistan is still much below the threshold level. This may not provide sufficient biodiversity and ecosystem services. The poor status of these two dimensions is expected to have a negative impact on food availability. This is particularly true in the case of Uzbekistan, where soil and water degradation are widespread. However, the trend lines of food supply are not in the expected lines. The reason

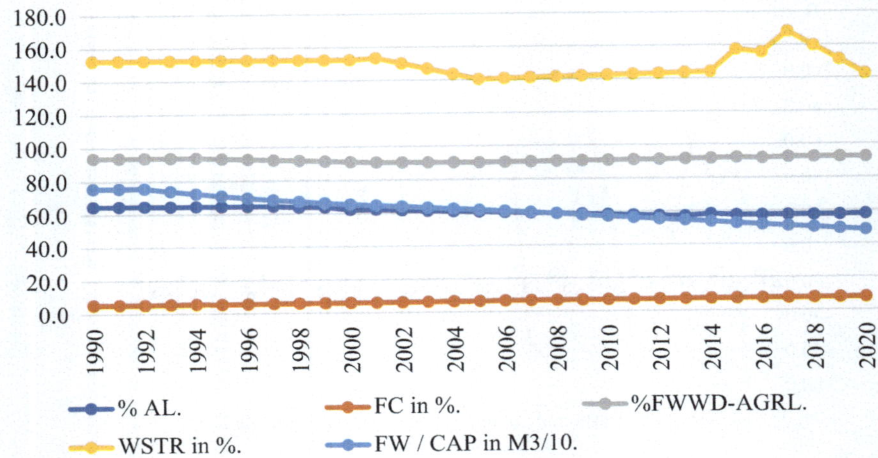

Fig. 2 Trajectories of nature indicators in Uzbekistan. *Source* https://ourworldindata.org/climate-change

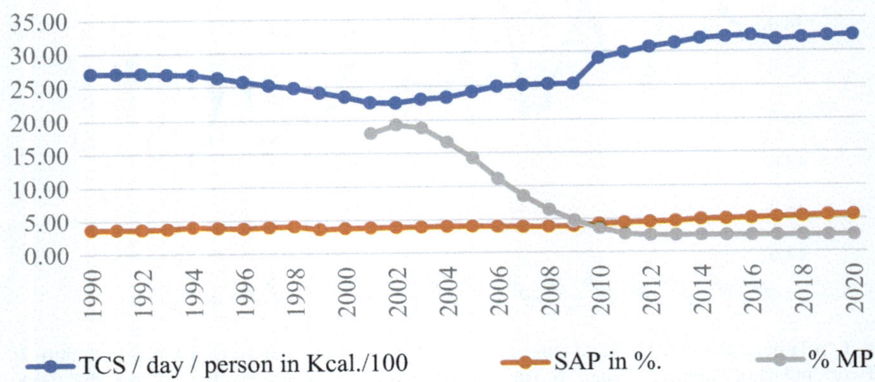

Fig. 3 Trajectories of food indicators in Uzbekistan. *Source* https://ourworldindata.org/climate-change

could be the intensification of agriculture, which is spreading at scale, coupled with a shift towards non-plant-based food. However, these trends are not sustainable in the current resource status and increasing climate risks, especially in the context of severe water stress. Thus, the impacts of the climate–nature–food nexus are yet to be fully established at the country level; however, they may be evident in certain regions of Uzbekistan, where air and water pollution are adversely impacting rural livelihoods (ADB 2012). Recognizing the adverse impacts associated with the C-N-F nexus, Uzbekistan is actively promoting afforestation on a massive scale.

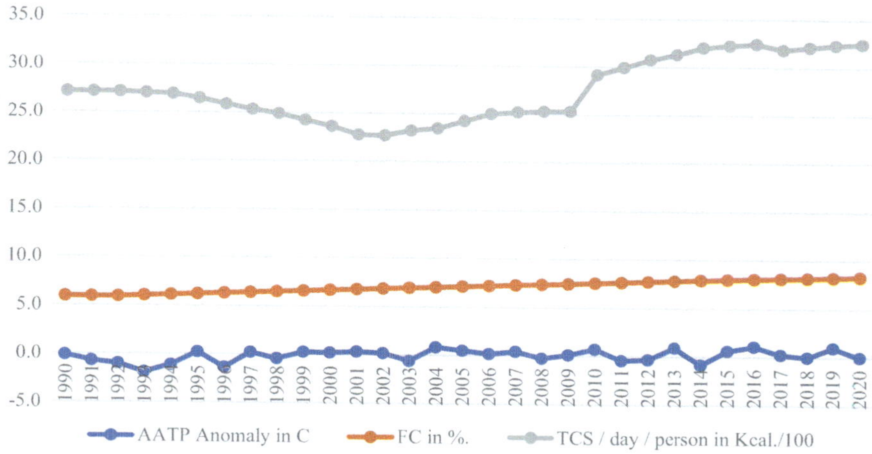

Fig. 4 Climate, nature, and food trajectories in Uzbekistan. *Source* https://ourworldindata.org/climate-change

3.2 C-N-F Nexus: Quantitative Assessment

Although Uzbekistan has a higher proportion of agricultural land compared to Central Asia, it faces twice the water stress of the region. Agricultural intensification is higher in Uzbekistan, with higher fertilizer application, as it consumes more than 90% of its available water resources for agricultural purposes. While Uzbekistan is one of the highly stressed countries in terms of natural endowments and climate risks, this is not reflected in the food dimension. These dimensions are assessed quantitatively in this section using the correlation coefficients. While the detailed correlation matrix is presented in the appendix (Table 12), this section only presents and discusses the significant correlation coefficients (Table 9). This section examines various indicators representing the three dimensions and their linkages with one another over the four periods. It is interesting to note that, despite the extreme conditions, the linkages among the important indicators were clearly evident.

3.3 Climate

Over the 30 year period, temperature anomalies, a main indicator of climate, have revealed a positive association with GDP, NPK use, forest cover, and the share of animal protein (Table 9). All these relationships align with theoretical expectations. The positive association between temperature anomaly and GDP and NPK use indicates that higher GDP levels and farm intensification increase climate risks due to an increase in GHG emissions. Increased consumption of animal protein is indirectly linked to climate change due to the reason that it imposes excessive

Table 9 Significant correlation coefficients of the C-N-F nexus indicators for Uzbekistan

Period	Climate—temperature	Nature—% forest cover	Food—total calorie supply/capita/day (kcal)
P1 (1990–2020)	% MP: –0.40***; SAP in %: 0.36***; RH in %. –0.45**; % AL: –0.53**; FC in %: 0.54**; FW/CAP in M^3. –0.55**: % FWWD-AGRL: –0.50**; NPK use in kgs/ha: 0.46***; GDP/capita/year in US$: 0.41***	TCS/day/person in kcal: 0.68*; % MP: –0.92*; SAP in %: 0.86*; RH in %: –0.74*; % AL: –0.97*; AATP anomaly in C: 0.54**; FW/CAP in M^3:1.00*; % FWWD-AGRL: –0.41**; NPK use in kgs/ha: 0.83*; GDP/capita/year in US$: 0.91	% MP: –0.81*; SAP in %: 0.87*; RH in %: –0.55*; % AL: –0.63**; FC in %: 0.68*; FW/CAP in M^3: –0.64*; % FWWD-AGRL: 0.35***; NPK use in kgs/ha: 0.77*; GDP/capita/year in US$: 0.88*
P2 (1990–2000)	TCS/day/person in kcal: –0.57**; RH in %: –0.66*; % AL: –0.43***; FC in %: 0.54**; FW/CAP in M^3. –0.51: % FWWD-AGRL: –0.59**; NPK use in kgs/ha: –0.59**	TCS/day/person in kcal: –0.98; RH in %: –0.59**; % AL: –0.80*; AATP anomaly in C: 0.54**; FW/CAP in M^3: –0.99*; % FWWD-AGRL: –0.97*; nitrogen use in kgs/ha: 0.65*; GDP/capita/year in US$: –0.54**	RH in %: 0.57**; % AL: 0.86*; FC in %: –0.98*; AATP anomaly in C: –0.57; FW/CAP in M^3: 0.95*; % FWWD-AGRL: 1.00*; nitrogen use in kgs/ha: 0.72*
P3 (2001–2010)	RH in %: –0.44***; SAP in %: 0.51**; potash use in kgs/ha: 0.49	TCS/day/person in kcal: 0.89*;% MP: –0.97*; SAP in %: 0.50***; % AL: –1.00*; FW/CAP in M^3: –1.00*; % FWWD-AGRL: 0.59**; WSTR in %: –0.78*; nitrogen use in kgs/ha: 0.78*; GDP/capita/year in US$: 0.99*	% MP: –0.88*; SAP in %: 0.75*; % AL: –0.88*; FC in %: 0.89*; AATP anomaly in C: 0.40***; FW/CAP in M^3: –0.90*; % FWWD-AGRL: 0.67*; WSTR in %: –0.59**; NPK use in kgs/ha: 0.80*; GDP/capita/year in US$: 0.91
P4 (2011–2020)	TCS/day/person in kcal: 0.44***; RH in %: –0.60**; % AL: –0.69*; AARF anomaly in mm: 0.67*; NPK: 0.48; GDP/capita/year in US$: 0.40***	TCS/day/person in kcal: 0.86*; % MP: –0.57**; SAP in %: 0.99*; FW/CAP in M^3: –1.00*; % FWWD-AGRL: 0.92*; WSTR in %: 0.46***; NPK use in kgs/ha: 0.73*; GDP/capita/year in US$: 1.00*	% MP: –0.76*; SAP in %: 0.85*; FC in %: 0.86*; AARF anomaly: 0.46***; AATP anomaly in C: 0.44***; FW/CAP in M^3: –0.82*; % FWWD-AGRL: 0.80*; WSTR in %: 0.43***; NPK use in kgs/ha: 0.76*; GDP/capita/year in US$: 0.86*

Note *, **, and *** indicate the level of significance at 1, 5, and 10% levels, respectively
Source Estimated by the authors

pressure on resources compared to plant protein. The positive association between temperature anomaly and forest cover in the case of Uzbekistan could be due to the below-threshold level of forest cover that is increasing, albeit marginally, in recent years. Temperature anomalies have revealed a negative association with the extent of malnourishment, percentage of agricultural land, water availability, and share of water use for agriculture, as well as relative humidity. All these are in perfect sync with climate risks. The temperature rise is expected to impact water resources directly, which in turn affects crop areas and production, i.e., the availability of food and nutrition. These linkages appear to be conspicuous and more consistent in the context of low resource endowments and high climate risks. All these relationships, except in the case of total calorie supply, are consistent over the time periods. Total calorie supply revealed a negative association with temperature anomalies during period 2 (1990–2000), but the coefficient turned positive during the last period (2011–2020). The positive sign is in line with the sub-regional assessment (Central Asia).

3.4 *Nature*

The percentage of forest cover, representing nature, has moved in the same direction as temperature anomalies, water availability, NPK use, GDP, total calorie supply, and the share of animal protein over the 30 year period (Table 9). Again, the positive association between NPK use, GDP, and temperature anomaly is not in support of the expected signs. Whereas, it has been moving in the opposite direction, with a decrease in the percentage of agricultural land, a decrease in the share of agricultural water use, an increase in the extent of malnutrition, and a decrease in relative humidity over the same period. The negative association between forest cover and agricultural water use indicates that agriculture is unable to draw enough water, despite improved water availability. This could be due to the extreme water stress conditions in Uzbekistan. This is also reflected in the negative association between forest cover and agriculture land. All these relationships are consistent in all the periods, except in some cases. For instance, the linkages between forest cover and water availability have turned out to be negative in all three sub-periods. Whereas the share of agriculture water use, which showed a negative association during the first (1990–2020) and second (1990–2000) periods, revealed a positive association during periods 3 (2001–2010) and 4 (2011–2020). And total calorie supply and GDP have revealed negative signs only during the second period (1990–2000), while they are positively associated with forest cover during the other three periods.

3.5 *Food*

Food is a crucial part of the nexus, and countries try to secure food, regardless of their natural and climatic conditions. Political entrepreneurs and policymakers tend to ensure food security by any means, especially in countries facing extreme conditions, such as Uzbekistan. Often, policy priorities facilitate food systems to overcome and perform beyond the natural constraints. As a result, variables such as NPK use (intensification) and GDP (socioeconomic development) turned out to be significant with positive signs and strengths (Table 9). Often, NPK use is complemented by irrigation water. The share of agricultural water turned out to be significant, with a positive sign, despite a negative association between water availability and total calorie supply. Forest cover and the share of animal protein have turned out to be significant with positive signs. The extent of malnutrition has a positive sign, indicating that calorie supply or food security may not ensure nutritional security. The share of agricultural land has turned out to be significant, with a negative sign, during the 30 year period. Though not expected, the percentage of AL turned positively significant during the second period. Otherwise, all the variables have been consistent over the periods. In fact, climate indicators like rainfall anomaly and temperature anomaly turned out to be significant with a positive sign.

3.6 C-N-F Nexus in the Context Poorly Endowed Backward Country

Uzbekistan stands out compared to the other countries, with the weakest economy, the most constrained natural resource base, and the harshest climatic conditions. The share of agricultural land has turned out to be negative, while forest cover revealed a positive association with total calorie availability. This indicates an increasing reliance on non-crop calories, as reflected in the strong correlation between total calorie availability and the share of animal protein (Table 10).

The correlations of freshwater withdrawals for agriculture and water stress are positive, stemming from the lower baseline levels of water availability and rainfall (Table 10). These circumstances also mean that intensification practices have been critical to improvements in calorie availability, both from the perspective of being a prerequisite for agriculture, and as a greening effect from increased crop cover in a rather desert-like biome. Despite this, the increased share of agricultural land still has a strong negative correlation with total calorie availability, while pesticide and fertilizer application has strong correlations with both forest cover and total calorie availability. However, in contrast to Nepal (Chap. 6), agriculture is prioritized higher than the conservation of water resources, as evidenced by the significantly higher positive correlation between water stress and freshwater withdrawals for agriculture in Uzbekistan.

The focus of Uzbekistan's agricultural growth is also trained squarely on food security, in contrast to Nepal's focus on improving fodder availability (Table 11). This is to say that, despite similar correlations and relationships between the C-N-F nexus vertices, the pathways for each of these correlations are vastly different, with significantly worse baseline levels of natural resources, food security, and rainfall.

Table 10 Correlation coefficients between food indicators and other indicators for Uzbekistan (1990–2020)

Indicator	TCS/day/person in kcal	SAP in %	% MP
SAP in %	0.87*	1	
% MP	–0.81*	–0.80*	1
RH in %	–0.55**	–0.66*	0.71*
% AL	–0.63**	–0.76*	0.93*
FC in %	0.68*	0.86*	–0.92*
% FWWD-AGRL	0.35***	0.05	0.19
WSTR in %	0.21***	0.21***	0.15
Nitrogen use in kgs/ha	0.59**	0.78*	–0.77*
Potash use in kgs/ha	0.77*	0.73*	–0.50**
Phosphate use in kgs/ha	0.57**	0.22***	–0.36***
NPK use in kgs/ha	0.77*	0.83*	–0.82*

Note *, **, and *** indicate the level of significance at 1, 5, and 10% levels, respectively
Source Estimated by the authors

It is interesting to note the overlap in these trends, despite the vast differences in outcomes—it establishes that similar to Japan (Chap. 5), given different baseline circumstances, the same activity can either degrade existing biodiversity and natural resources base, or contribute to the greenification and alleviation of harsh climatic conditions in regions like Uzbekistan, establishing that no strategy can be objectively considered helpful or harmful agnostic of prevailing circumstances, as evidenced by the diverging forest cover data.

Table 11 Correlation coefficients between food indicators and other indicators for Uzbekistan and Nepal (1990–2020)

	% FWWD-AGRL		FC in %	
	Uzbekistan	Nepal	Uzbekistan	Nepal
WSTR in %	0.56**	–1.0	–0.12	–0.95*
Pesticide use in kgs/ha	–	–	–	0.82*
Nitrogen use in kgs/ha	–	–	0.87*	0.20
Potash use in kgs/ha	–	–	0.43**	0.46**
Phosphate use in kgs/ha	–	–	0.11	0.28***
NPK use in kgs/ha	–	–	0.83*	0.24***

Note *, **, and *** indicate the level of significance at 1, 5, and 10% levels, respectively
Source Estimated by the authors

4 Conclusions

The context of Central Asia and Uzbekistan presents an extreme situation in terms of natural endowments, climate risks, economic strengths, and policy environment. Under these circumstances, the trends and linkages between C-N-F indicators clearly establish that the indicators of climate and nature complement each other in a way that degradation of one element can adversely influence the other. Given that Central Asia and Uzbekistan are facing natural resource degradation and climate risks, the country-level situation is significantly more severe than the regional-level situation. While this clearly establishes a nexus between climate and nature, the joint influence of climate and nature is not being reflected in the food dimension. This is true of all the sub-regions and the case study countries. The reason being that a proactive policy environment has helped in overcoming the C-N-F nexus effectively. This is particularly evident in the context of Uzbekistan, which reported a higher total calorie supply compared to other case study countries, including Japan. While national food policies ensure food availability, environmental policies appear to be cognizant of the fact that continuing such support is not sustainable and needs to be addressed in a sustained manner. Keeping this in view, Uzbekistan has embarked on a massive program of afforestation to improve biodiversity and ecosystem services. While Nepal is in a comparable situation, its approach seems to be different. Nepal's path is to strengthen the C-N-F nexus, i.e., maintaining and enhancing economic growth at the cost of environmental degradation and natural capital, as it can rely on its better natural resource endowments.

Appendix

See Table 12.

Table 12 Correlation coefficients of the C-N-F nexus indicators for Uzbekistan

Period/Indicator	1990–1920			1990–2000			2001–2010			2011–2020		
	TCS/day/person in kcal	FC in %	AATP anomaly in C	TCS/day/person in kcal	FC in %	AATP anomaly in C	TCS/day/person in kcal	FC in %	AATP anomaly in C	TCS/day/person in kcal	FC in %	AATP anomaly in C
TCS/day/person in kcal	1.00	0.68*	0.15	1.00	–0.98*	–0.57**	1.00*	0.89*	0.31	1.00*	0.86*	0.44***
SAP in %	0.87*	0.86*	0.36***	–0.14	0.29	0.04	0.75*	0.50***	0.51*	0.85*	0.99*	0.32
% MP	–0.81*	–0.92*	–0.40***	–	–	–	–0.88*	–0.97*	–0.16	–0.76*	–0.57**	–0.35
RH in %	–0.55**	–0.74*	–0.45**	–	–	–0.66*	–0.28	–0.25	–0.44***	0.00	–0.22	0.60**
% AL	–0.63**	–0.97*	–0.53**	0.86*	–0.80*	–0.43***	–0.88*	–1.00*	–0.15	0.02	0.02	–0.69*
FC in %	0.68*	1.00	0.54**	–0.98*	1.00	0.54**	0.89*	1.00	0.16	0.86*	1.00	0.37
AARF anomaly in mm	–0.17	–0.24	–0.16	0.35	–0.39	–0.38	0.40***	–0.35	–0.34	0.46***	0.18	0.67*
AATP anomaly in C	0.15	0.54**	1.00	–0.57**	0.54**	1.00	0.31	0.16	1.00	0.44***	0.37	1.00
% FWWD-AGRL	0.35***	–0.41***	–0.50**	1.00*	–0.97*	–0.59**	0.67***	0.59**	0.03	0.80*	0.92*	0.35
WSTR in %	0.21	–0.12	–0.08	0.00	0.00	0.00	–0.59**	–0.78*	–0.17	0.43***	0.46***	0.35
Nitrogen use in kgs/ha	0.59**	0.87*	0.58**	–0.72*	0.65*	0.66**	0.75*	0.70*	–0.13	0.72*	0.83*	0.39
Potash use in kgs/ha	0.77*	0.43***	0.11	0.70*	–0.65*	–0.48***	0.65*	0.66*	0.49***	0.69*	0.76*	0.50**
Phosphate use in kgs/ha	0.57**	0.11	–0.19	0.69*	–0.81*	–0.48***	0.75*	0.77*	0.11	–0.03	–0.38	0.03

(continued)

Table 12 (continued)

Period/Indicator	1990–1920			1990–2000			2001–2010			2011–2020		
	TCS/day/ person in kcal	FC in %	AATP anomaly in C	TCS/day/ person in kcal	FC in %	AATP anomaly in C	TCS/day/ person in kcal	FC in %	AATP anomaly in C	TCS/day/ person in kcal	FC in %	AATP anomaly in C
NPK use in kgs/ ha	0.77*	0.83*	0.46***	−0.38	0.26	0.44**	0.80*	0.78*	0.02	0.76*	0.73*	0.48***
FW/CAP in M^3	−0.64*	−1.00*	−0.55**	0.95*	−0.99*	−0.51**	−0.90*	−1.00*	−0.16	−0.82*	−1.00*	−0.34
GDP/capita/ year in US$	0.88*	0.91*	0.41***	0.39	−0.54**	0.03	0.91*	0.99*	0.17	0.86*	1.00*	0.40***

Note *, **, and *** indicate the level of significance at 1, 5, and 10% levels, respectively
Source Estimated by the authors

References

ADB (2012);Country strategy and programme: Uzbekistan (2006–2010); Asian Development Bank. https://www.adb.org/sites/default/files/institutional-document/32403/csp2006-2010_0.pdf

Artem L, Ryazantsev S, Moiseeva E, Manshin R (2020) The economic and social consequences of environmental migration in the central Asian countries. Central Asia Caucasus 21(2). https://doi.org/10.37178/ca-c.20.2.13

IRC; LNRMI; CESS and WASSAN (2013) Sustainable water and sanitation services: the life-cycle cost approach to planning and management in India, Earth Scan/Routledge, UK.

Karimov Y, Musaev I, Mirzababayeva S, Abobakirova Z, Umarov S, Mirzaeva Z (2023) Land use and land cover change dynamics of Uzbekistan: a review. In: E3S web of conferences, vol 421, p 03007 SERBEMA-2023. https://doi.org/10.1051/e3sconf/202342103007

Lioubimtseva E, Henebry GM (2009) Climate and environmental change in arid Central Asia: impacts, vulnerability, and adaptations. J Arid Environ 73(11):963–977. https://doi.org/10.1016/j.jaridenv.2009.04.022

Reddy VR Rahut DB, Sonobe T (2025) Future of small holder farms in Asia. Asian Development Bank Institute, Tokyo, Japan. https://www.adb.org/sites/default/files/publication/1041036/future-smallholder-farming-asia.pdf

Chapter 5
East Asia Case Study: Japan

1 Introduction

East Asia is among the high-performing regions of the world, with Japan, China, and South Korea leading the pack. These three countries, along with Mongolia, are the only four countries in the region, as per the taxonomy of the ADB. The economic performance of the region is driven by manufacturing, with China being a global manufacturing hub. As a result, the region faces severe environmental problems, including air and water pollution (Yang et al. 2020). At the same time, the region is endowed with more varied biomes, ranging from verdant forests to snow-capped mountains and temperate island climates. East Asia has rich biodiversity and diverse socioeconomic systems and outcomes, thus presenting an interesting perspective for assessing the C-N-F nexus. What trade-offs are made in the name of food availability, industrial development, and growing housing needs? Japan and South Korea, meanwhile, are at the forefront of Asia's technological growth. These two countries have transitioned smoothly from agriculture to the industry and service sectors, while China is in the process, and Mongolia is still in the primary stage. Since Mongolia is a small nation, the regional context is influenced by China on one extreme and Japan and Korea on the other. Being robust economies (all three) based on manufacturing and intellectual property development, they provide an interesting insight into how economies with lower levels of socioeconomic stress (and recovery from world-changing events, as in the case of Japan) fare in the C-N-F nexus. The region with more varied biomes is a departure from the arid geographies of Central Asia.

East Asia is characterized by high inter-country variations, especially in terms of environmental status. These variations are attributed to the differences in the nature and type of manufactured goods and the production processes adopted, such as efficient technologies and environmentally friendly technologies. While Japan ranks 20th globally in the environmental performance index, China ranks 120th (Huang and Xu 2019). Japan has a high score on the environmental health index, while China has a low score. In terms of ecosystem vitality, China has a high score, which may

V. R. Reddy et al., *Strengthening the Climate–Nature–Food Nexus*,
https://doi.org/10.1007/978-3-032-29423-4_5

be attributed to its vast and diverse biodiversity. Being the global manufacturing hub and center of technological advancement, the associated environmental problems are spilling beyond the region and contributing globally as well (Tominaga 2001). Given the high levels of economic performance and increasing environmental concerns, East Asia would provide interesting insights into the C-N-F nexus.

2 Quantitative Assessment of C-N-F Nexus: East Asia

East Asia is characterized by an average annual rainfall of above 1000 mm, along with high humidity levels (64.4%), and shows a positive rainfall anomaly (Table 1). The average annual temperature anomaly is showing little change, although studies observe that East Asia is experiencing a rise in temperatures (You et al. 2022). East Asia is endowed with a good forest cover (40.9%) and a relatively low proportion (34.4%) of area under agriculture. This could be due to high industrialization as well as urbanization in the region. Given the good rainfall and forest cover, water availability is quite high, and water stress levels are low. The proportion of total water used for agriculture is as high as 61%, despite the fact that only a third of the land is allocated to agriculture. Crop intensification is also reflected in the use of chemical fertilizers and pesticides per unit of land. East Asia has a high per-capita GDP, a reasonably good calorie supply, and a low extent of malnutrition. The region is also reporting low and declining poverty rates in recent years.[1] Thus, East Asia is among the best-performing regions in the world, with a good backup of natural capital and vibrant ecosystems. Despite these positives, the region faces severe climate risks that threaten the sustainability of socioeconomic development. This is attributed to the unsustainable production practices. A more detailed quantitative assessment is conducted using regional-level correlation coefficients. The number of indicators representing the three dimensions has turned out to be significant in all four periods, although the direction and strength of the linkages vary between the periods (Table 2). Despite the intra-regional variations and being the better-endowed and performing region, the signs of the coefficients not only endorse the C-N-F nexus observed in the sub-regional visualization (Chap. 2) and the quantitative assessment at the aggregate (Asia) level. Indicators representing each of the C-N-F dimensions and their interlinkages across the four periods are discussed below.

2.1 *Climate*

All the natural indicators, viz., forest cover, water availability, and share of agricultural land, are moving in the same direction as temperature anomalies over the past

[1] https://www.worldbank.org/en/region/eap/overview.

Table 1 Descriptive statistics of selected indicators for East Asia

Indicator	C-N-F	Descriptive statistics	
		Mean	SD
AARF in mm	Climate	1026.6	91
AARF anomaly in mm	Climate	4.04	85.6
AATP anomaly in C	Climate	0	0.4
RH in %	Climate	64.4	1.6
% FWWD-AGRL	Nature	61.3	3.4
WSTR in %	Nature	41.9	0.3
% AL	Nature	34.4	1.4
FC in %	Nature	40.9	0.3
FW/CAP in M^3	Nature	5131.2	509
TCS/day/person in kcal	Food	2867	119
TPS/capita/day in grams	Food	90.8	10.2
APS/capita/day in grams	Food	45.6	7.4
SAP in %	Food	6.7	0.7
% MP	Food	8.4	3.2
GDP/capita/year in US$	Economic	18,663	4111
Nitrogen use in kgs/ha	Intensification	116.3	7.4
Phosphate use in kgs/ha	Intensification	76.1	14.1
Potash use in kgs/ha	Intensification	60.3	11.6
NPK use in kgs/ha	Intensification	252.7	30.7
Pesticide use in kgs/ha	Intensification	7.1	0.6

30 year period (Table 2). This could be due to the fact that natural capital indicators have been moving positively over the years, along with a marginal change in climate (temperature anomaly). Along with nature indicators, total calorie supply, GDP per capita, and NPK use have turned out to be positive. All these linkages are on the expected lines. Only relative humidity and extent of malnutrition move in the opposite direction to temperature anomalies (Table 2). While the negative association between temperature and relative humidity is expected, the negative association with the extent of malnutrition could be due to changes in the microenvironment. Increasing air and water pollution is widespread in these regions (Yang et al. 2020). While sub-periods 2 (1990–2000) and 3 (2001–2010) have revealed only a few indicators with significance, some important indicators turned up significant with opposite signs when compared to the 30 year period (1990–2020). That is, water availability and the share of agricultural land turned out to be strong and significant, with a negative sign. Water stress also turned out to be significantly negatively in the more recent decade. This indicates changing C-N-F nexus linkages over the years.

Table 2 Significant correlation coefficients of the C-N-F nexus indicators for East Asia sub-region

Period	Climate—temperature	Nature—% forest cover	Food—total calorie supply/capita/day (kcal)
P1 (1990–2020)	RH in %: –0.27***; NPK use in kgs/ha: 0.32***; FC in %: 0.51**; FW/CAP in M^3: 0.55**; % AL: 0.40***; GDP/capita/year in US$: 0.52**; TCS/day/person in kcal: 0.54**; %MP: –0.46***	FW/CAP in M^3: 0.99*; TCS/day/person in kcal: 0.96*; %MP: –0.95*; RH in %: –0.74*; NPK use in kgs/ha: –0.69*; WSTR in %: 0.79*; pesticide use in kgs/ha: 0.79*;% AL: –0.94*; SAP in %: 0.80*; AATP anomaly in C: 0.51**	AATP anomaly in C: 0.54**; RH in %: –0.69*; FC in %: 0.96*; FW/CAP in M^3: –0.97*; NPK use in kgs/ha: –0.69*; WSTR in %: 0.67*; pesticide use in kgs/ha: –0.82*; % AL: –0.89*; %MP: –0.96*
P2 (1990–2000)	% AL: 0.50**	FW/CAP in M^3: –0.99*; NPK use in Kgs/ha: 0.63**; pesticide use in Kgs/ha: 0.86*; GDP/capita/year in US$: 0.96*; SAP in %: 0.88*; TCS/day/person in kcal: 0.43***; RH in %: –0.55**	FC in %: 0.43***; FW/CAP in M^3: –0.45***
P3 (2001–2010)	AARF anomaly in mm: –0.52**	FW/CAP in M^3: –1.00*; NPK use in kgs/ha: –0.62**; WSTR in %: 0.92*; pesticide use in kgs/ha: –0.82*; % AL: –0.82*; GDP/capita/year in US$: 0.97*; SAP in %: 0.65*; TCS/day/person in kcal: 0.98*; %MP: –0.98*; RH in %: –0.51**	RH in %: –0.58**; FC in %: 0.98*; FW/CAP in M^3: –0.98*; NPK use in kgs/ha: –0.63**; pesticide use in kgs/ha: –0.74*; % AL: –0.71*; WSTR: 0.89*; %MP: –0.99*
P4 (2011–2020)	FW/CAP in M^3: –0.85*; FC in %. 0.82*;WSTR in %: –0.59**; GDP: 0.84*; TCS/day/person in kcal: 0.75*;%MP: –0.86*; % AL: –0.75*	FW/CAP in M^3: –1.00*; NPK use in kgs/ha: –0.57**; WSTR in %: 0.77*; GDP/capita/year in US$: 0.93*; SAP in %: 0.98*; % AL: –0.98*; TCS/day/person in kcal: 0.97*; %MP: –0.99*; AATP anomaly in C: 0.82*	AATP anomaly in C: 0.75*; FC in %: 0.97*; FW/CAP in M^3: 0.96*; NPK use in kgs/ha: –0.55*; WSTR in %: –0.73*; % AL: –0.94*; %MP: –0.95*

Note *, **, and *** indicate the level of significance at 1, 5, and 10% levels, respectively
Source Estimated by the authors

2.2 Nature

The prime indicator of nature (forest cover) is positively associated with water availability, calorie supply, water stress, pesticide use, the share of animal protein, and temperature anomaly (Table 2). In the case of water stress, revealing a positive association with forest cover is due to the reason that the demand for water seems to be outstripping the availability of quality water. Aggressive industrialization is adversely impacting water resources. Even the water availability turned negative and was strongly significant in all the sub-periods, vindicating the demand–supply gap. Given the relatively high coverage of forest in the region, coupled with a low proportion of agricultural land, there is a high trade-off between forest cover and agricultural land use. While NPK use has shown a negative association with forest cover in the longer period, though its sign is not consistent. The same is the case with pesticide use. However, these two indicators are not directly related to forest cover. The share of animal protein has revealed a positive association with forest cover, and the relationship is strong and consistent across the periods. This suggests that forest cover and animal protein supply are mutually complementary.

2.3 *Food*

Food is a critical element of the C-N-F nexus, irrespective of the region's socioeconomic and environmental status. Interestingly, the expected direct linkages to total calorie supply, such as water, agricultural land, fertilizer, and pesticide use, have shown a consistent negative association (Table 2). This is an anomaly that needs further probing. Only forest cover and temperature anomalies have shown a positive association with total calorie supplies. The positive association between temperature anomalies and total calorie supply is similar to that observed in the case of Asia. And the negative relationship between total calorie supply and the extent of malnutrition, which supports the earlier observations in the case of Asia. It may be noted that GDP per capita turned out to be significant and consistent with a positive sign in most cases. However, this does not seem to have any significance in the case of East Asia. However, some of the counterintuitive relations are discussed in the following section.

2.4 *High-Performing Industrial Economies and the C-N-F Nexus*

East Asia comprises some of the world's strongest economies, including Japan, South Korea, and China. While each of these economies has grown strong through different pathways, they share significant strengths in terms of climatic stability, the abundance of natural resources, and strong food security and availability. Each of the countries provides insight into how managing the C-N-F nexus well can result in positive outcomes across all three facets.

With more varied climatic zones within these nations, the correlations will be harder to ascertain at the micro-level than with Central Asia; however, it is encouraging that the broad trends still corroborate the overall effects within the C-N-F, as illustrated by the Asia Aggregate data. The East Asian sub-regional analysis further exposes the pitfalls of aggregate analysis in general—smaller nations (geographically or economically) tend to be overshadowed by larger datasets, leading to improper attribution of correlations. In this section, the analysis focuses on more intricate linkages and a comparison of East Asia with the Asia aggregates. The East Asia assessment will be largely attributed to the outcomes for China due to its overwhelming size; therefore, more analysis must be conducted at the national level for Japan/Korea.

While the correlation coefficients differ slightly, total calorie supply, temperature anomalies, and relative humidity all align with the Asia averages, indicating a general overall trend among these variables. The outlier, however, is in the correlation between forest cover and calorie availability (Table 3). The data for East Asia is indicative of calorie sources from forests, given the high correlation between the two variables, paired with a highly negative correlation between the share of agricultural land and calorie availability. This is further corroborated by the highly negative correlations between the intensification variables (use of nitrogen, phosphate, potash, pesticide use, and anomalous correlation with water stress). This is also the case with the last decadal data (Table 4). While it may suggest that the relationship between the C-N-F factors is counter-indicative for this region, the remaining indicators do not support this hypothesis, and as such, these anomalous findings need to be examined in greater detail.

Table 3 Correlation coefficients of food indicators with other indicators pertaining to Asia and East Asia for the period 1990–2020

Indicators	Asia		East Asia	
	TCS/day/person in kcal	% MP	TCS/day/person in kcal	% MP
FC in %	–0.42***	0.36***	0.96*	–0.95*
% FWWD-AGRL	–0.24	0.13	–0.37***	0.25***
FW/CAP in M^3	–0.83*	0.78*	–0.97*	0.94*
Nitrogen use in kgs/ha	0.93*	–0.89*	–0.69*	0.74*
Phosphate use in kgs/ha	–0.09	0.12	–0.69*	0.76*
Potash use in kgs/ha	0.08	–0.06	–0.51***	0.51**
NPK use in kgs/ha	0.75*	–0.71*	–0.69*	0.73*
WSTR in %	–0.89*	0.94*	0.67*	–0.68*
Pesticide use in kgs/ha	–0.14	0.22***	–0.82*	0.90*
% AL	0.54**	–0.47**	–0.89*	0.90*

Note *, **, and *** indicate the level of significance at 1, 5, and 10% levels, respectively
Source Estimated by the authors

Table 4 Correlation coefficients of food indicators with other indicators pertaining to Asia and East Asia for the period 2011–2020

Indicators	Asia		East Asia	
	TCS/day/person in Kcal	% MP	TCS/day/person in kcal	% MP
FC in %	–0.99*	0.97*	0.97*	–0.99*
FW/CAP in M^3	–0.99*	0.97*	–0.96*	1.00*
Nitrogen use in kgs/ha	0.92*	–0.85*	–0.51**	0.37
Phosphate use in kgs/ha	0.67*	–0.78*	–0.71*	0.71*
Potash use in kgs/ha	0.59**	–0.71*	–0.24	0.18
NPK use in kgs/ha	0.92*	–0.91*	–0.55**	0.48***
WSTR in %	0.15	–0.31	–0.73*	0.73*
Pesticide use in kgs/ha	0.69*	–0.42***	–0.21	0.06
% AL	0.96*	–0.82*	–0.94*	0.95*

Note *, **, and *** indicate the level of significance at 1, 5, and 10% levels, respectively
Source Estimated by the authors

This is especially noteworthy in the case of China, where intensification has significantly contributed to alleviating critical food security challenges following periods of famine. The data thus do not indicate a negative effect in this case, but rather a low attribution of the availability as a direct result of these factors. This may primarily be linked to a historical reliance on non-agrarian sources of nutrition and calories, such as meat, seafood, and forest systems. This is further supported by the high correlation between protein availability and forest cover (r = 0.92). It also sets the foundation for the increasingly corrosive impact of manufacturing activity in China, where loss of forest cover (Chap. 2) largely does not contribute to growth in agrarian area, but rather a conversion into industrial manufacturing, contributing poorly to calorie availability, despite providing a balancing effect through improved per-capita GDP within the country. China, due to its size (especially in comparison to the small island nations of Japan and South Korea), results in an over-indexation of the sub-region aggregate towards the outcomes within China, explaining the high coefficients (both positive and negative).

The over-indexation towards China's outcomes is also evident in the intercorrelation between the intensification variables. The correlations for nitrogen application to forests are strongly negatively correlated, similar to those in Asian aggregates. However, the correlation for the Asian aggregates was supported by an expanding agricultural area resulting from the conversion of forest cover to agricultural land; the data for East Asia, however, do not show the same correlations. The data for nitrogen, phosphate, and potash application showed a strong positive correlation with calorie availability, whereas for East Asia, these coefficients were negative. Finally, the context of water stress variable and its correlation to other intensification indicators shows the outcomes being influenced by externalities (such as industrial expansion). It also further corroborates that calorie availability comes with water

Table 5 Correlation coefficients of food indicators with other indicators pertaining to East Asia for the period 1990–2020

Indicator	WSTR in %
Nitrogen use in kgs/ha	–0.59*
Phosphate use in kgs/ha	–0.56*
Potash use in kgs/ha	–0.58*
NPK use in kgs/ha	–0.62*

Note *, **, and *** indicate the level of significance at 1, 5, and 10% levels, respectively
Source Estimated by the authors

stress, which is especially pertinent in outcomes with a higher focus on non-agrarian sources of nutrition and calories (meat and seafood). This would mean that there should be a high negative correlation between the soil support variables (N, P, K) and calorie availability, as these measures tend to increase water stress and thus have a net negative impact on animal husbandry and pisciculture. However, this is not the case for East Asia. In fact, even pesticide use is negatively correlated with water stress (–0.82), showing that the impact of agricultural intensification and expansion is dwarfed by the deteriorative effect of industrial activity (Table 5).

Thus, the story for China (and, consequently, this limited macro-analysis for East Asia) is one of animal husbandry and industrial activity applying pressure on water systems, resulting in a pseudo-resilience at the macro-level to climatic pressures so far. However, as admitted earlier, this is far too generalized to draw conclusive evidence from, and therefore a deeper country-level analysis must be conducted to truly understand the learnings from the C-N-F nexus and its significance in Eastern Asia.

While the aggregate indicators for the region are more representative of the Chinese context, given the wealth of research focused on the China case study, we find that it is more valuable to conduct a more incisive analysis of the Japanese data to draw meaningful insights on economies with a well-managed C-N-F nexus.

3 C-N-F Nexus: Case of Japan

Japan is the first country in Asia to have transformed from an underdeveloped agricultural economy into an industrial and then a technology- and service-based economy. For the last four decades, Japan has been competing with the global West in terms of economic performance and well-being, despite facing disaster during the Second World War. Japan is the only Asian country that has found a place in the G-5 nations, i.e., the present-day G-20. Along with economic prosperity, Japan is facing environmental problems, including GHG emissions and air and water pollution. In the context of agriculture, Japan is suffering from a declining and aging population. Japan is adopting sustainable agriculture practices, including reducing chemical use, paying farmers for environmentally friendly practices, and educating them about good farming methods (Fuhrmann-Aoyagi et al. 2024). In fact, Japan is one of the most food-insecure countries in Asia, as it imports 60% of its calorie intake. These sustainable practices or measures are not enough, as Japan's environmental problems are more closely associated with energy, construction, and transportation. For coal continues to be the main source of GHG emissions in Japan, as it accounts for 30% of power generation.[2] Although Japan is at the top of Asia in terms of environmental performance, environmental health, and ecosystem vitality, it is showing a declining trend (Huang and Xu 2019). To address these concerns, Japan's public policy promotes energy efficiency, green energy, green building, and green transport initiatives, along with sustainable farming practices.

Within East Asia, Japan is well endowed with natural resources, such as rainfall and forest cover. Rainfall anomaly is positive while temperature anomaly is zero in terms of deviation from the 30 year averages (Table 6). Forest cover accounts for as much as 68% of the geographical area. As a result, the share of agricultural land is very low at 14%. This could be the reason for Japan's food security issues in recent years. On the other hand, per-capita freshwater availability in Japan is substantially lower when compared to East Asia, although water stress is marginally lower in Japan. Japan also uses a higher share (66%) of its freshwater for agricultural purposes. The use of fertilizer in Japan is lower, although farmers seem to use higher doses of pesticides. The decline in chemical usage could be attributed to recent policies promoting sustainable agriculture. This is reflected in the marginally lower calorie supply, accompanied by a higher animal protein supply. The extent of malnutrition is substantially lower in Japan compared to the East Asian average. This indicates that Japan's food basket is nutritionally dense. The economic prosperity of Japan is well reflected in its substantially higher per-capita GDP. Some of the differences between East Asia and Japan are mainly due to the overshadowing impact of China.

3.1 C-N-F Nexus: Trends

While the previous section discussed the 30 year averages of various indicators, this section examines the trends of the important indicators of C-N-F over the 30 year period. Trends in climate indicators show that temperature anomalies and relative humidity have remained stable over the 30 year period (Fig. 1). In contrast, rainfall anomalies have exhibited high fluctuations with an increasing trend. It is worth noting that rainfall anomalies have been predominantly positive for most years since 2008, with only 2 years showing a marginal decline. Therefore, it is clear that Japan is experiencing increasing rainfall anomalies and stable temperature and relative humidity anomalies.

In the case of the natural dimension, most indicators revealed a stable trend over the period (Fig. 2). Only the share of agricultural land has shown some decline.

[2] https://www.oecd.org/en/publications/2025/03/oecd-environmental-performance-reviews-japan-2025_947dc3da.html.

Table 6 Descriptive statistics of the selected C-N-F indicators for Japan

Indicator	C-N-F	Descriptive statistics	
		Mean	SD
AARF in mm	Climate	1768	155
AARF anomaly in mm	Climate	1.7	155
AATP anomaly in C	Climate	0	0.4
RH in %	Climate	72.5	1.1
% FWWD-AGRL	Nature	66	1.3
WSTR in %	Nature	38.9	2.2
% AL	Nature	14.2	0.9
FC in %	Nature	68.4	0.1
FW/CAP in M^3	Nature	3394	34.3
TCS/day/person in kcal	Food	2793	97.9
SAP in %	Food	7.4	0.2
% MP	Food	2.6	0.2
GDP/capita/year in US$	Economic	32,429	2190
Nitrogen use in kgs/ha	Intensification	76.4	9.6
Phosphate use in kgs/ha	Intensification	100.6	23.4
Potash use in kgs/ha	Intensification	67.6	13.3
NPK use in kgs/ha	Intensification	244.6	45.2
Pesticide use in kgs/ha	Intensification	13.8	1.9

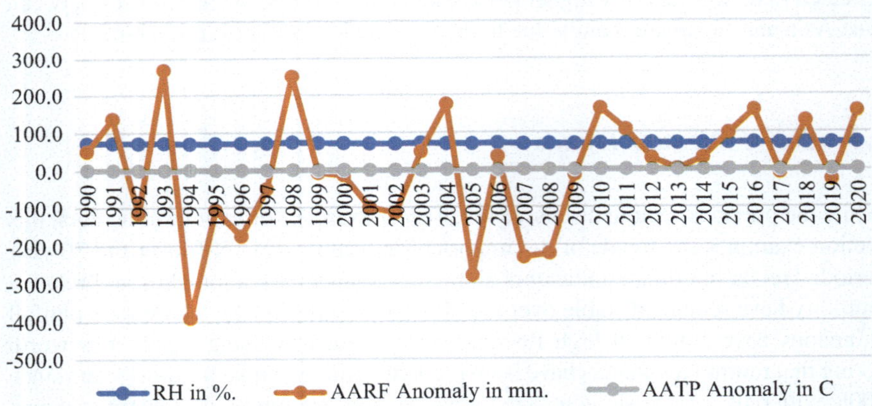

Fig. 1 Trajectories of rainfall and temperature anomalies (climate) in Japan. *Source* https://ourworldindata.org/climate-change

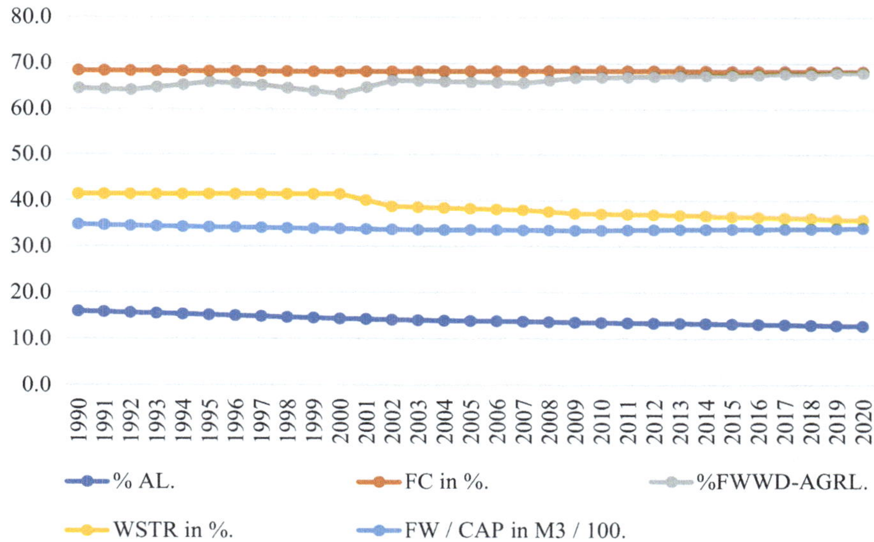

Fig. 2 Trajectories of nature indicators in Japan. *Source* https://ourworldindata.org/climate-change

This is due to the increasing demand for urbanization and industrial use. While water availability has remained constant, the share of water used for agriculture has increased by about 3%. Water stress has decreased from approximately 42% to about 34% over the period. This could be due to the substantial decline in the area under agriculture.

Trajectories of food indicators have revealed interesting trends. The total calorie supply has declined from approximately 2900 kcal to about 2700 kcal over the past 30 years (Fig. 3). There is a marginal increase in the share of animal protein in the supply. Moreover, the extent of malnutrition has increased, though marginally. This is unexpected in the case of Japan, given its economic prosperity. Although Japan's per-capita GDP has seen a moderate increase over the period, its average per-capita GDP is quite high and unlikely to adversely impact food supplies. However, it has been observed that Japan is facing food security issues lately due to declining crop lands and a decreasing number of farmers growing crops. However, this may not significantly impact food supply or availability, as Japan is well equipped to import food at any cost. One explanation could be the aging population, i.e., food consumption is expected to decline with age. In any case, total calorie availability is much higher than the requirement (Fig. 4).

The core indicators of the C-N-F dimensions reveal flat trends, as they move in parallel over the period. While forest cover remained constant, temperature anomalies have shown a marginal increase. This endorses the C-N-F nexus hypothesis, though in a moderate manner. Such subdued linkages could be due to the presence of natural capital in Japan. In fact, a better status of biodiversity (forest and water resources) could be the reason for the mild trends in climate change. Although direct evidence

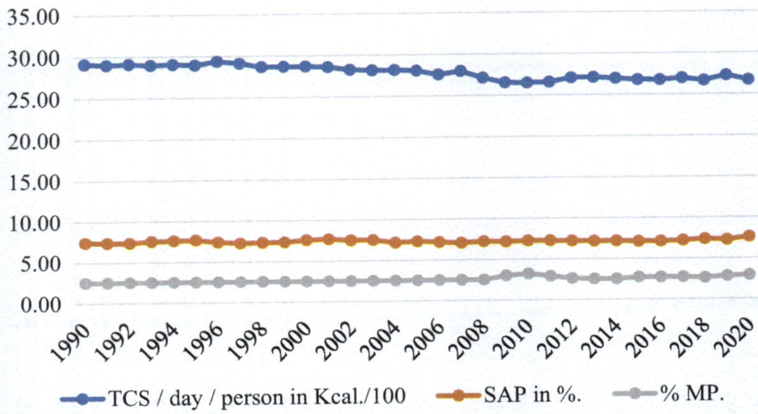

Fig. 3 Trajectories of food indicators in Japan. *Source* https://ourworldindata.org/climate-change

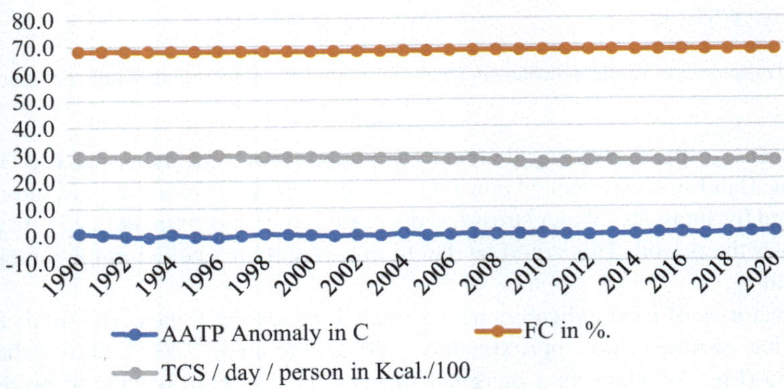

Fig. 4 Climate, nature, and food trajectories in Japan. *Source* https://ourworldindata.org/climate-change

of climate change, in terms of temperature and rainfall, is less conspicuous, the indirect impacts of climate change (natural disasters, air and water pollution) are very evident. Even these indirect impacts are being addressed effectively in Japan. The quantitative assessment presented in the next section may provide more insights into this.

3.2 C-N-F Nexus: Quantitative Assessment

Japan assumes a unique place in our assessment of the C-N-F nexus. It is an island nation with abundant natural resources, socioeconomic potential, and technological strengths. In terms of economic development, it is on par with the Western developed countries of Europe and the Americas. In this regard, it would be interesting to examine and understand how the C-N-F nexus is evolving. While the trend analysis has already provided some interesting insights, this section quantitatively assesses the linkages. The assessment is carried out between and within each of the C-N-F dimensions, covering four periods: the long-term period of 30 years (1990–2020) and the three sub-periods, which cover the three consecutive decades (1990–2000, 2001–2010, and 2011–2020). The number of indicators that turned out to be significant in all three dimensions. While the detailed correlation matrix is presented in the appendix (Table 10), only the significant correlation coefficients are presented and discussed here (Table 7).

3.3 Climate

Temperature anomalies, representing the climate dimension, are positively associated with GDP, forest cover, rainfall anomalies, agricultural water use, and the extent of malnutrition over a 30 year period (Table 7). Climate (temperature) and nature (forest) are moving in the same direction. This could be due to the fact that in the case of Japan, these two indicators have moved within a narrow margin over the period. Even the positive association between temperature and rainfall anomalies could be due to the same reason. Whereas the positive relationship between GDP, agricultural water use, and the extent of malnutrition is on the expected lines. Agriculture needs more water as the average temperature rises. Similarly, the negative association between temperature anomalies, on the one hand, and total calorie availability, share of agricultural land, water stress, pesticide use, NPK use, and share of animal protein, on the other, is as expected. For agricultural intensification associated with chemical use (NPK and pesticides), the trend moves in the opposite direction to temperature anomalies. And hence the negative association with agricultural land and water stress. Most of these indicators exhibit consistent signs over different periods. Only forest cover has shown a negative sign in the recent decade. This indicates that recent years have seen the strengthening of the climate–nature nexus.

3.4 Nature

Forest cover (nature) is positively associated with the extent of malnutrition, temperature anomalies, agriculture water use, and GDP per capita (Table 7). On the other hand, forest cover has a negative association with relative humidity, the share of animal protein, the share of agriculture land, water stress, and chemical use (NPK and pesticide use). While the share of agriculture land and chemical use is expected to move in the opposite direction from forest cover, the share of animal protein and water stress is likely to move in the same direction. Such anomalies are observed

Table 7 Significant correlation coefficients of the C-N-F nexus indicators for Japan

Period	Climate—temperature	Nature—% forest cover	Food—total calorie supply/capita/day (Kcal)
P1 (1990–2020)	TCS/day/person in kcal: –0.46**; SAP in %: –0.34***; % MP: 0.39***; % AL: –0.43**; FC in %: 0.30***; AARF anomaly in mm: 0.28***; % FWWD-AGRL: 0.38**; WSTR in %: –0.46**; pesticide use in Kgs/ha: –0.47**; NPK use in kgs/ha: –0.44**; GDP/capita/year in US$: 0.42**	TCS/day/person in Kcal: –0.66*; SAP in %: –0.54**; % MP: 0.54**; RH in %: –0.40***; % AL: –0.35; AATP anomaly in C: 0.30***; % FWWD-AGRL: 0.60**; WSTR in %: –0.61**; pesticide use in Kgs/ha: –0.76*; NPK use in kgs/ha: –0.44**; GDP/capita/year in US$: 0.38***	SAP in %: 0.53**;% MP: –0.69*; RH in %: 0.36***; % AL: 0.90*; FC in %. –0.66*; AARF anomaly in mm: –0.29***; AATP anomaly in C: –0.46**; % FWWD-AGRL: –0.85*; WSTR in %: 0.94*; pesticide use in kgs/ha: 0.93*; NPK use in kgs/ha: 0.93*; FW/CAP in M^3: 0.66*; GDP/capita/year in US$: –0.82*
P2 (1990–2000)	TCS/day/person in kcal: –0.42***	TCS/day/person in Kcal: 0.53**; % AL: 1.00*; pesticide use in Kgs/ha: –1.00*; NPK use in kgs/ha: 0.89*; FW/CAP in M^3: 0.99*; GDP/capita/year in US$: –0.91*	RH in %: –0.47***; % AL: 0.54**; FC in %: 0.53**; AARF anomaly in mm: –0.44***; AATP anomaly in C: –0.42***; % FWWD-AGRL: 0.58**; pesticide use in kgs/ha: –0.54**; NPK use in kgs/ha: 0.48***; FW/CAP in M^3: 0.50**
P3 (2001–2010)	TCS/day/person in kcal: –0.46***; SAP in %: –0.75*; RH in %: –0.42***; % AL: –0.59**; FC in %: 0.58**; AARF anomaly in mm: 0.42***; % FWWD-AGRL: 0.51**; WSTR in %: –0.62**; pesticide use in kg/ha: –0.68*; FW/CAP in M^3: –0.63**	TCS/day/person in kcal: –0.93*; SAP in %: –0.73*; % MP: 0.73*; % AL: –1.00*; AATP anomaly in C: 0.58**; % FWWD-AGRL: 0.70*; WSTR in %: –0.93*; pesticide use in Kgs/ha: 0.90*; NPK use in kgs/ha: –0.64*; FW/CAP in M^3: –0.92*; GDP/capita/year in US$: 0.54*	SAP in %: 0.51**; % MP: –0.83*; % AL: 0.91*; FC in %: –0.93*; AATP anomaly in C: –0.46***; % FWWD-AGRL: –0.80*; WSTR in %: 0.88*; pesticide use in kgs/ha: 0.77*; NPK use in kgs/ha: 0.83*; FW/CAP in M^3: 0.79*

(continued)

Table 7 (continued)

Period	Climate—temperature	Nature—% forest cover	Food—total calorie supply/capita/day (Kcal)
P4 (2011–2020)	% MP: 0.40***; RH in %: 0.77*; % AL: –0.77*; FC in %: –0.79*; AARF anomaly in mm: 0.47***; % FWWD-AGRL: 0.73*; WSTR in %: –0.76*; NPK use in kgs/ha: –0.92*; FW/CAP in M^3: 0.79*; GDP/capita/year in US$: 0.61**	SAP in %: –0.41*; RH in %: –0.69; % AL: 0.94*; AATP anomaly in C: –0.79*; % FWWD-AGRL: –0.95*; WSTR in %: 0.94*; NPK use in kgs/ha: 0.77*; FW/CAP in M^3: –0.96*; GDP/capita/year in US$: –0.87*	% MP: –0.40***; AARF anomaly in mm: –0.87*; pesticide use in kgs/ha: 0.49***

Note *, **, and *** indicate the level of significance at 1, 5, and 10% levels, respectively
Source Estimated by the authors

even in the sub-periods. The signs of important indicators are not consistent over the years. This indicates the changing C-N-F nexus linkages over the periods, especially during the recent decade, and also due to some indirect linkages between different indicators. These indirect linkages will be explored later.

3.5 Food

The food system is directly influenced by climate and nature indicators. Over the 30 year period, the supply/availability of calories is positively associated with the share of agricultural land, water stress, NPK use, pesticide use, the share of animal protein, and relative humidity (Table 7). All these relationships are theoretically valid—except in the case of water stress. Whereas agriculture water use, rainfall and temperature anomalies, the extent of malnutrition, and GDP per capita have revealed a negative association with calorie supply. While the negative association of the extent of malnutrition is logical, the negative association with other indicators of nature can be explained as agriculture is not heavily influenced by marginal variations in rainfall, temperature anomalies, and water due to favorable rainfall conditions. The negative association between GDP and calorie availability could be due to the demographic structure (aging population). Most of the indicators have shown a consistent association across the sub-periods, although some inconsistencies are evident in certain sub-periods. Some of the anomalies observed across the three dimensions will be explored further by analyzing the interlinkages among the indicators.

3.6 Innovation/Technology-Led Developed Economy

The geographic and socioeconomic foundations of Japan yield some interesting learnings on how to manage the C-N-F nexus. Great leaps in technology and manufacturing have fueled Japan's economic growth. What is most important to understand is that this has been achieved while placing a great emphasis on preserving nature and keeping tight controls on pollution and ecological degradation. It may be argued that Japan has maintained its ecological base without much trading off with rapid economic growth. Japan is a testament to the fact that the conservation of nature and prudent utilization of natural resources do not need to come at the cost of economic prosperity. Japan follows similar trends (although with different outcomes) as the Asia aggregate. There is an emphasis on expanded agricultural output, followed by intensification (Table 8). In the process, it maintained a stable and healthy share of animal and seafood protein in the calorie composition.

It can be said that Japan may be the closest to serving as an example of near-ideal management of the C-N-F nexus. However, over time, Japan's demographic makeup has skewed in a direction that further magnifies the effects of a more conservative approach, with lower calorie availability levels today than at the beginning of the 30 year period. This might suggest that food security has declined over the period under review, but this is likely the result of an aging population, paired with an economy that is progressively assigning higher weight to the manufacturing and service industries.

Table 8 Correlation coefficients of specific C-N-F nexus indicators for Japan for 1990–2020 and 2011–2020

Indicator	1990–2020			2011–2020		
	TCS/day/person in kcal	SAP in %	% MP	TCS/day/person in kcal	SAP in %	% MP
SAP in %	0.53**	1		–0.34	1	
% MP	–0.69*	–0.21***	1	–0.40***	0.51**	1
% AL	0.90*	0.48**	–0.51*	0.01	–0.65*	–0.41***
WSTR in %	0.94*	0.54**	–	–0.10	–0.60**	–
Pesticide use in Kgs/ha	0.93*	0.58**	–	0.49***	–0.53**	–
Nitrogen use in kgs/ha	0.89*	0.52**	–	0.001	–0.23	–
Potash use in kgs/ha	0.87*	0.35***	–	0.32	–0.55**	–
Phosphate use in kgs/ha	0.94*	0.54**	–	–0.23	–0.33	–
NPK use in kgs/ha	0.93*	0.49***	–	0.15	–0.50	–

Note *, **, and *** indicate the level of significance at 1, 5, and 10% levels, respectively
Source Estimated by the authors

Table 9 Correlation coefficients of specific C-N-F nexus indicators for Japan for 1990–2020

	TCS/ day/ person in kcal	SAP in %	% MP	RH in %	% AL	FC in %	AARF anomaly in mm	AATP anomaly in C
% AL	0.90*	0.48**	–0.51**	0.33***	1			
FC in %	–0.66*	–0.54**	0.54**	–0.40***	–0.35***	1		
AARF anomaly in mm	–0.29***	–0.14	0.32***	0.54**	–0.16	0.17	1	
AATP anomaly in C	–0.46**	–0.34***	0.39***	0.03	–0.43***	0.30***	0.28***	1
% FWWD-AGRL	–0.85*	–0.43**	0.56**	–0.37***	–0.84*	0.60**	0.16	0.38***

Note *, **, and *** indicate the level of significance at 1, 5, and 10% levels, respectively
Source Estimated by the authors

For this reason, it is difficult to directly attribute the findings from the last decade as indicative of the overall C-N-F nexus (Table 8). However, it may be important to recognize the impact that demographic shifts may have on the outcomes of initiatives or strategies that previously resulted in such strong positive impacts.

Japan's approach to agriculture is particularly interesting when considered in the context of the rest of Asia. While there has been both expansion and intensification, both of which correlate with higher levels of calorie availability, coupled with the application of fertilizers and pesticides, it would certainly appear that the pathways for Japan mirror those of the rest of Asia (Table 9). However, the forest cover correlations, as well as the anomaly data, show that this expansion and intensification have been applied with precision, revealing an emphasis on scientific agriculture, rather than an aggressive approach to resource exploitation. It is essential to draw key lessons from Japan's example—economic prosperity does not necessarily come from the exploitation of natural resources. Conservative monetary policy, coupled with a focus on advancing indigenous technology, creates opportunities to safeguard natural balances and resources while addressing the needs of the population.

Japan's approach to agriculture is particularly interesting when considered in the context of the rest of Asia. While there has been both expansion and intensification, both of which correlate with higher levels of calorie availability, coupled with the application of fertilizers and pesticides, it would certainly appear that the pathways for Japan mirror those of the rest of Asia (Table 9). However, the forest cover correlations, as well as the anomaly data, show that this expansion and intensification have been applied with precision, revealing an emphasis on scientific agriculture, rather than an aggressive approach to resource exploitation. It is essential to draw key lessons from Japan's example—economic prosperity does not necessarily come from the exploitation of natural resources. Conservative monetary policy, coupled with a focus on advancing indigenous technology, creates opportunities to safeguard natural balances and resources while addressing the needs of the population.

4 Conclusions

As it is often argued, in the context of most developed countries, growth/development and degradation often go hand in hand. Some exceptions to this argument are the Scandinavian countries, which are effectively balancing growth and prosperity. In the Asian context, Japan is setting such an example of effectively managing the C-N-F nexus. This could be due to the high policy priorities set for environmental protection and development and adoption of environmental-friendly technologies across the board, including agriculture. Although these strategies are slowing down growth and even impacting food production, they are paying dividends in the form of preserving biodiversity and ecosystem services. These strategies range from promoting sustainable farm practices and payments for ecosystem services to farmers to the promotion of green and sustainable technologies. While the number of countries in Asia has similar policies, most of them remain on paper, and very few are implemented. However, Japan could implement all these policies on the ground, and the results are evident. Japan provides important insights into the C-N-F nexus and learnings for systematically managing the C-N-F.

Appendix

See Table 10.

Table 10 Correlation coefficients of the C-N-F nexus indicators for Japan

Indicator	1990–1920			1990–2000			2001–2010			2011–2020		
	TCS/day/ person in Kcal	FC in %	AATP anomaly in C	TCS/day/ person in Kcal.2	FC in %	AATP anomaly in C4	TCS/day/ person in Kcal.5	FC in %	AATP anomaly in C7	TCS/day/ person in Kcal.8	FC in %	AATP anomaly in C10
TCS/day/person in Kcal	1.00	–0.66*	–0.46**	1.00	0.53**	–0.42**	1.00	–0.93*	–0.46***	1.00	–0.04	–0.19
SAP in %	0.53**	–0.54**	–0.34***	–0.07	0.05	–0.21	0.51**	–0.73*	–0.75*	–0.34	–0.41***	0.38
% MP	–0.69*	0.54**	0.39***	–	–	–	–0.83*	0.70*	0.39	–0.40***	–0.19	0.40
RH in %	0.36***	–0.40**	0.03	–0.47***	–0.24	0.22	0.21	–0.38	–0.42***	–0.26	–0.69*	0.77*
% AL	0.90*	–0.35***	–0.43**	0.54**	1.00	0.04	0.91*	–1.00*	–0.59**	0.01	0.94*	–0.77*
FC in %	–0.66*	1.00*	0.30***	0.53**	1.00	0.07	–0.93*	1.00	0.58**	–0.04	1.00	–0.79*
AARF anomaly in mm	–0.29***	0.17	0.28***	–0.44***	0.03	0.05	–0.25	0.09	0.42***	–0.87*	–0.15	0.47***
AATP anomaly in C	–0.46**	0.30***	1.00	–0.42***	0.07	1.00	–0.46***	0.58**	1.00	–0.19	–0.79*	1.00
% FWWD-AGRL	–0.85*	0.60**	0.38***	0.58**	0.19	–0.37	–0.80*	0.70*	0.51**	0.14	–0.95*	0.73*
WSTR in %	0.94*	–0.61**	–0.46**	–	–	–	0.88*	–0.93*	–0.62**	–0.10	0.94*	–0.76*
Pesticide use in kgs/ha	0.93*	–0.76*	–0.47**	–0.54**	–1.00*	–0.05	0.77*	–0.90*	–0.68*	0.49***	0.08	–0.28
Nitrogen use in kgs/ha	0.89*	–0.44**	–0.40**	0.40***	0.87*	0.04	0.84*	–0.71*	–0.26	0.00	0.36	–0.70*
Potash use in kgs/ha	0.87*	–0.30***	–0.46**	0.60**	0.91*	–0.09	0.68*	–0.38	–0.07	0.32	0.76*	–0.83*

(continued)

Table 10 (continued)

Indicator	1990–1920			1990–2000			2001–2010			2011–2020		
	TCS/day/ person in Kcal	FC in %	AATP anomaly in C	TCS/day/ person in Kcal.2	FC in %	AATP anomaly in C4	TCS/day/ person in Kcal.5	FC in %	AATP anomaly in C7	TCS/day/ person in Kcal.8	FC in %	AATP anomaly in C10
Phosphate use in kgs/ha	0.94*	–0.50**	–0.43**	0.38	0.79*	–0.13	0.84*	–0.70*	–0.26	–0.23	0.73*	–0.81*
NPK use in kgs/ ha	0.93*	–0.44**	–0.44**	0.48***	0.89*	–0.08	0.83*	–0.64*	–0.21	0.15	0.77*	–0.92*
FW/CAP in M3	0.66*	–0.10	–0.12	0.50**	0.99*	0.13	0.79*	–0.92*	–0.63**	0.00	–0.96*	0.79*
GDP/capita/ year in US$	–0.82*	0.38***	0.42**	–0.24	–0.91*	–0.27	–0.24	0.54**	0.39	0.19	–0.87*	0.61**

Note *, **, and *** indicate the level of significance at 1, 5, and 10% levels, respectively
Source Estimated by the authors

References

Fuhrmann-Aoyagi MB, Miura K, Watanabe K (2024) Sustainability in Japan's agriculture: an analysis of current approaches. Sustainability 16:596. https://doi.org/10.3390/su16020596

Huang B, Xu Y (2019) Environmental performance in Asia: overview, drivers, and policy implications; ADBI Working Paper 990. Tokyo: Asian Development Bank Institute. https://www.adb.org/publications/environmental-performance-asia-overview-drivers-policy-implications

Tominaga K (2001) East Asia: environmental issues. In: Smelser NJ, Baltes PB (eds) International encyclopedia of the social and behavioral sciences. Pergamon, pp 3943–3947. https://doi.org/10.1016/B0-08-043076-7/03265-4

Yang L, Wang Y, Wang R et al (2020) Environmental-social-economic footprints of consumption and trade in the Asia-Pacific region. Nat Commun 11:4490. https://doi.org/10.1038/s41467-020-18338-3

You Q, Jiang Z, Yue X et al (2022) Recent frontiers of climate changes in East Asia at global warming of 1.5 and 2 °C. Climate Atmos Sci 5:80. https://doi.org/10.1038/s41612-022-00303-0

References

Chapter 6
South Asia Case Study: Nepal

1 Introduction

South Asia is among the most socioeconomically and environmentally vulnerable regions in the world. The region accounts for the highest number of people living below the poverty line, and its cities are among the most polluted in the world. The region is among the high-performing in terms of GDP growth rates and reports a sharp decline in poverty levels. But the region is relatively slow in its economic transformation, unlike East Asia. More than 50% of the population in the region continues to depend on agriculture and operates on suboptimal farm sizes. This region is known for its intensive agricultural practices, particularly in terms of water- and chemical-intensive methods. As a result, land degradation, deforestation, loss of biodiversity, water scarcity, water pollution, and both indoor and outdoor air pollution are widespread, adversely impacting ecosystem services. Additionally, natural resource degradation (soil, water, and biodiversity) is impacting crop yields in some regions and causing severe health problems (Reddy et al. 2025). A changing climate is further aggravating the vulnerabilities and livelihoods of the rural population. It is observed that approximately 750 million people in South Asia are affected by climate risks, which are manifested in heatwaves, cyclones, droughts, and floods. Furthermore, the region is a victim of growth-induced environmental problems associated with affluence and inequities. It also faces the adverse environmental impacts associated with the transfer of poor technologies from developed countries. Within Asia, most of its member countries are ranked very high, even above those in Central Asia, in terms of environmental performance (Huang and Xu 2019).

The region presents a good ground for understanding the linkages between the C-N-F nexus, as it is characterized by low and declining natural resources (forests, water, soils, etc.) coupled with high growth and vulnerable food security. At the same time, climate risks are accentuated due to a low resource base and the degradation of natural resources and biodiversity. In fact, resource degradation is threatening food security and self-sufficiency in some countries. However, the region is progressing well, at

V. R. Reddy et al., *Strengthening the Climate–Nature–Food Nexus*,
https://doi.org/10.1007/978-3-032-29423-4_6

least in the policy front, in terms of adopting environmentally friendly technologies such as climate-smart agricultural practices, regenerative forests, solar and wind energy, and hydro power (Lal n.d.). However, the region lags behind in implementing policies and strategies at the grassroots level. As in the case of East Asia, India has an overwhelming influence on the region's averages due to its geographical and demographic size; hence, the regional assessment is mostly representative of and influenced by the Indian situation. India, being a dominant player in the region, has relatively high intra-regional variations. Its geographical location in South Asia falls between Central Asia and the rest of Asia in terms of biodiversity and climatic conditions.

2 Quantitative Assessment of C-N-F Nexus: South Asia

South Asia, comprising Bangladesh, Bhutan, India, Nepal, and Sri Lanka, is often referred to as an emerging economy with high growth performance, particularly among its member countries, such as India and Bangladesh. The aggregate picture of South Asia is somewhat misleading, especially regarding its natural resource endowments, as the regional picture may not accurately reflect or represent all the countries in the region due to the wide variations between them. Bhutan is a notable outlier, particularly in terms of precipitation, forest cover, and water resources. Bhutan is among the top five countries in the world in terms of these three resources, and hence it affects the regional averages.

At the aggregate level, South Asia has a high average rainfall of above 2000 mm and the highest freshwater availability at 27,730 M3, mainly due to Bhutan's water situation, which is among the top 5 countries in the world (Table 1). Average annual rainfall anomalies have shown an increase of 5 mm over the period of 30 years. However, temperature anomalies have been reported as unchanged. South Asia has a moderate level of forest cover (36.6%), although this level is largely due to the high forest cover (approximately 70%) in Bhutan. Without Bhutan, the forest cover in South Asia would be below 30%. South Asia has a relatively high proportion of area under agriculture. Agriculture accounts for 92% of the total water available in the region, indicating the high dependence on agriculture and irrigation. Despite high water availability, water stress remains significant due to a substantial demand–supply gap in countries like India. Crop intensification is at a moderate level in the region, although some countries, such as Bangladesh and India, have achieved high intensification levels in terms of fertilizer (NPK) use per hectare. Although pesticide use per hectare is very low at the regional level, countries like India, Bangladesh, and Sri Lanka are reporting a high incidence of pesticide-linked health problems (Reddy et al. 2025). The region is characterized by resource use inefficiencies and low productivity. This is reflected in the moderate level of food availability or calorie supply, which is on the margin of per capita requirement, in the region. Food composition is biased towards carbs, as proteins account for only 11% of the total calories. South Asia consumes the least amount of animal protein (1.9%).

Table 1 Descriptive statistics of selected indicators for South Asia

Indicator	C-N-F	Descriptive statistics	
		Mean	SD
AARF in mm	Climate	2088.2	152.9
AARF Anomaly in mm	Climate	5	152.9
AATP Anomaly in C	Climate	0	0.4
RH in %	Climate	75.3	1
%FWWD-AGRL	Nature	92.3	0.9
WSTR in %	Nature	46.1	10.1
% AL	Nature	43	0.5
FC in %	Nature	36.6	0.5
FW/CAP in M^3	Nature	27,730	4515
TCS/day/person in Kcal	Food	2363.6	190.1
TPS/capita/day in grams	Food	58.2	8.2
APS/capita/day in grams	Food	11.5	3.3
SAP in %	Food	1.9	0.4
% MP	Food	14.8	4.1
GDP/Capita/Year in US$	Economic	1427.2	579.8
Nitrogen use in Kgs/ha	Intensification	61.9	12.4
Phosphate use in Kgs/ha	Intensification	21.5	6.7
Potash use in Kgs/ha	Intensification	14.7	3.3
NPK use in Kgs/ha	Intensification	98.1	20.7
Pesticide use in Kgs/ha	Intensification	0.5	0.2

The region's food habits are environmentally friendly, as they depend mostly on plant-based proteins. And malnutrition is quite high at about 14% despite declining poverty levels (hunger).

While South Asia is termed an emerging economic region, it has a lot to catch up on in terms of food availability, resource management, and other areas. Despite high growth performance in recent decades, the region continues to have low GDP per capita. South Asia is also among the climate hotspots and is expected to face climate risks, including increasing temperatures, glacier melt, and air and water pollution. Part of this is due to the unsustainable crop practices, polluting industries (being a dumping ground for dirty technologies), high dependence on imports, etc. The interlinkages of various indicators representing the C-N-F dimensions are explored in detail, using correlation coefficients, at the regional level over thirty years, as well as for the three sub-periods. The number of indicators representing the three dimensions has turned out to be significant in all four periods, although the direction and strength of the linkages vary over the periods (Table 2).

Table 2 Significant correlation coefficients of the C-N-F nexus indicators for South Asia Sub-region

Period	Climate—Temperature	Nature—% Forest Cover	Food—Total Calorie Supply/capita/day (Kcal)
P1 (1990–2020)	FW/CAP in M^3): –0.67; FC in %: 0.66; NPK use in kgs./ha.: 0.55; WSTR in %: –0.56; Pesticide use in Kgs/ha.: 0.60; % AL: 0.46; GDP/capita/year in US$: 0.62; TCS/day/person in Kcal: 0.60;%MP: –0.63	FW/CAP in M^3: –0.96; WSTR in %: –0.88; GDP/capita/year in US$:0.97; SAP in %: 0.89; TCS/day/person in Kcal: 0.94; %MP: –0.96; AATP anomaly in C: 0.66; RH in %: 0.44	AATP anomaly in C: 0.60; RH in %: 0.50; FC in %: 0.94; FW/CAP in M^3: –0.87; NPK use in kgs./ha.:0.89; WSTR in %: –0.85; Pesticide use in Kgs/ha: 0.96; % AL: 0.70; % MP: –0.96; %FWWD-AGRL: 0.78; GDP/capita/year in US$: 0.98; TPS in grams: 0.99; SAP in %: 0.98
P2 (1990–2000)	%FWWD-AGRL: 0.42: NPK use in kgs/ha.: 0.49	FW/CAP in M^3: –0.88; NPK use in kgs/ha: 0.94; GDP/capita/year in US$: 1.00; SAP in %: 0.94; TCS/day/person in Kcal: 0.77	FC in %: 0.77; FW/CAP in M^3: –0.83; NPK use in kgs/ha.: 0.85; %FWWD-agrl: –0.47; GDP/capita/year in US$: 0.77; TPA in grams: 0.96; SAP in %: 0.67
P3 (2001–2010)	FW/CAP in M^3: –0.71; RH in %: –0.64; NPK use per ha.: 0.55; FC in %: 0.79; Pesticide use in kgs/ha: 0.80; WSTR in %: –0.64; GDP/capita/year in US$: 0.80; % AL: 0.45; TCS/day/person in Kcal: 0.76; %MP: –0.74	FW/CAP in M^3: –0.96; WSTR in %: 0.80; GDP/capita/year in US$: 0.99; SAP in %: 0.83; % FWWD-agrl: –0.63; TCS/day/person in Kcal: 0.97; % MP: –0.98; AARF mm: 0.42; AARF Anomaly in mm: 0.42:AATPAnomaly in C: 0.79	AATP anomaly in C: 0.76; FC in %:0.97; FW/CAP in M^3:-0.90; NPK use per ha.: 0.55; Pesticide use in kgs/ha: 0.92; WSTR in %: –0.86; AARF Anomaly in mm: 0.41; % MP: –0.99; AARF in mm: 0.41; AARF anomaly in mm:0.41; AATP Anomaly in C: 0.76; %FWWD-agrl: –0.92; GDP/capita/year in US$: 0.98; TPA in grams: 0.91; SAP in %: 0.81
P4 (2011–2020)	FW/CAP in M^3: –0.53; FC in %: 0.56; GDP/capita/year in US$: 0.58; %MP: –0.64	FW/CAP in M^3: –0.99; NPK use per ha: 0.60; GDP/capita/year in US$: 0.97; SAP in %: 0.84; % AL: 0.68; TCS/day/person in Kcal: 0.92; % MP: –0.97; AATP anomaly in: 0.56; RH in %: 0.69	AARF Anomaly in mm: 0.45; RH in %: 0.67; FC in %: 0.92; FW/CAP in M^3: –0.96; NPK use per ha.: 0.76; Pesticide use in kgs/ha: 0.85; % AL: 0.86; MP in %: –0.86; AARF in mm: 0.45; GDP/capita/year in US$: 0.90; TPA in grams: 1.00; SAP in %: 0.98

Note *, **, and *** indicate the level of significance at 1, 5, and 10% levels, respectively
Source Estimated by the authors

2.1 Climate

Nature indicators, such as forest cover and the share of agricultural land, are moving in the same direction as temperature anomalies over the past 30 year period (Table 2). While forest cover is expected to have an inverse association with temperature anomalies, the positive association revealed here could be due to the below-threshold level of forest cover in most countries (especially India) in the region. Whereas the positive sign of agricultural land is expected, as increased intensification and unsustainable farm practices tend to increase climate risks. This is also reflected in the positive association of intensification variables (NPK and pesticides). Similarly, the food dimension (total calorie availability) is also positively associated with temperature anomalies. Water availability and water stress have moved in opposite directions over the past 30 years. In fact, water stress is increasing along with temperature in several regions of India. GDP per capita has also turned out to be positive, as higher GDP levels often tend to increase temperature anomalies. And, the extent of malnutrition is negatively associated with temperature anomalies, which is expected. All these linkages and their signs are consistent across the three sub-periods, although fewer indicators turned out to be significant in two periods (1990–2000 and 2011–2020). Relative humidity has turned out to be significant with a negative sign during the period from 2001 to 2011.

2.2 Nature

Forest cover in South Asia is below the threshold level (33%), especially in India and Bangladesh, and has been receiving policy support over the past three decades. This has resulted in some improvement in the forest cover in these countries. It is positively associated with GDP, calorie supply, the share of animal protein, temperature anomalies, and relative humidity over a longer period (Table 2). Water availability and water stress became increasingly negative and strongly significant over the thirty years. This reflects the widening demand–supply gap, and forest cover is not helping to reduce this gap, given the below-threshold level of forest cover in large countries like India. In fact, the relationship between water availability is not only consistent across all sub-periods but also became stronger during the recent decade, vindicating the widening demand–supply gap. Given the low forest cover in countries like India, there appears to be no trade-off between forest cover and agricultural land use, as they move in the same direction (a positive relation during the recent decade). And NPK use has also shown a positive association with forest cover during the recent decade (2011–2020). The share of animal protein has revealed a positive association with forest cover, and the relationship is strong and consistent across the periods. This suggests that forest cover and animal protein supply are mutually complementary. And the climatic indicators of relative humidity and temperature have turned out to be significant with positive signs during the later periods (P3 and P4).

2.3 Food

As the region undergoes rapid economic progress, food availability is also directly influenced by the GDP per capita. All the climate indicators, viz., temperature, rainfall, and relative humidity, are positively associated with total calorie supply. Their relationship has been consistent and has grown stronger over time. The positive association between temperature anomalies and food supply supports (Mendelsohn's 2014) observation. While water use for agriculture proved significant, with an expected positive sign in the longer period (1990–2020), it turned negative in the subsequent two sub-periods (P2 and P3). Water availability also showed a negative association with calorie supply in all the periods. Land use in agriculture, along with the use of fertilizer and pesticides, has shown a consistent positive sign in all periods, especially during the 30 year period and the recent decade. The positive association between forest cover and total calorie supply indicates the complementarity between nature and food. And the negative relationship between total calorie supply and the extent of malnutrition, which supports the earlier observations in the case of Asia. The total protein supply and the share of animal protein are moving in line with the total calorie supply, which is on expected lines.

2.4 C-N-F Nexus in the Context of Developing Economies

South Asia presents interesting insights. Developing nations are working tirelessly to balance their natural resources from a sustainable development perspective and are also aggressively expanding their economies to participate in the global economy. This is especially true in the case of India, Bangladesh, and Sri Lanka, with some nations, such as Bhutan and Nepal, indexing more towards preservation than economic expansion. India and Bangladesh, in particular, have become key export hubs in the global economy. While India's primary export continues to be human resources, considerable trade-offs need to be constantly weighed against each other when it comes to developmental policy and ecological prudence. In Bangladesh's case, the textile industry has contributed significantly to the country's economic expansion, although this has come at a great environmental expense, particularly in terms of pollution and resource degradation.

Within the context of the C-N-F nexus interactions, externalities inevitably impact the interactions between the nexus indicators. Given the current state of understanding regarding the nexus itself, there is considerable value in understanding and evaluating the forces within the nexus, independent of externalities such as industrial activity and hard infrastructure development (e.g., dams, roads). To that end, South Asia does not present many surprises when contextualized against the Asian average. Once again, the geographic, climatic, and socio-economic circumstances inform some of the interactions between these variables and therefore color the findings as well. The story of South Asia is one of aggressive food security and self-sufficiency measures over the past 30 years and the externalities associated with these measures (unsustainable adoption of green revolution technologies, etc.).

These policies and outcomes are demonstrated in the correlation data for South Asia; the anomalous findings are explained by the historical policy-making and widespread adoption of intensification technologies in the region. The correlations for the application of nitrogen, phosphate, and potash are the highest among all the sub-regions, particularly in stark contrast to the Asian average (Table 3). Other indicators of increased caloric supply from animal protein are consistent with the rest of the sub-regions and the Asia average. The application of NPK fertilizers and pesticides drives the caloric growth. South Asia also demonstrates a similar trend of a high correlation between caloric availability and forest cover, but this can be understood within the context of Bhutan and Nepal, where agriculture is not conducted in the same manner as in India, Bangladesh, and Sri Lanka. Bhutan and Nepal place greater emphasis on preserving forest cover, and nutrition is derived from existing natural foliage and fauna.

The most interesting findings are related to water stress—while it may seem that high stress is correlated with higher availability, this would be a false attribution. Instead, this indicates that agriculture is being conducted even in high-water stress scenarios, aided by additional intensification technologies to maintain food security.

The overall trend across the three decades provides a clearer picture of the nature of the interactions between the variables. The latest decadal data is encouraging, however, as it illustrates that policy-making is moving in the right direction. The aggregate analysis for South Asia has a few outliers when compared to the Asian aggregate analysis. However, there is significantly more value in the drill-down analysis at the country level to better understand vulnerabilities within these nations and how trends across Asia are either nullified or magnified at the local level. These nations have also worked to diversify their solutions to food security, introducing new crop varieties and shifting diets (promotion of millets in India, for instance).

Table 3 Correlation coefficients between total calories supply and intensification and nature indicators for the periods 1990–2020 and 2001–2020

Indicator	1990–2020		2011–2020	
	Asia	South Asia	Asia	South Asia
Nitrogen use in Kgs/ha	0.93*	0.75*	0.92*	0.51**
Phosphate use in Kgs/ha	–0.09	0.94*	0.67*	0.91*
Potash use in Kgs/ha	0.08	0.90*	0.59*	0.87*
NPK use in Kgs/ha	0.75*	0.89*	0.92*	0.76*
WSTR in %	–0.89*	–0.85*	0.15	–0.00
Pesticide use in Kgs/ha	–0.14	0.96*	0.69*	0.85*
% AL	–	–	0.96*	0.86*

Note *, **, and *** indicate the level of significance at 1, 5, and 10% levels, respectively
Source Estimated by the authors

Improved incomes and globalization also significantly influenced the dietary composition and production mix of crops in these nations, and these trends will be studied in more detail at the country-level analysis.

3 C-N-F Nexus: Case of Nepal

Nepal is a landlocked small Himalayan country. It is among the poorest countries in the world (ranked at 205), and it is food insecure despite the policy focus on self-sufficiency (Knerr 2017). Over the past decade, Nepal has made significant progress, achieving decent GDP growth and attaining lower-middle-income status by 2020. Poverty has declined substantially between 2014 (30%) and 2019 (17%) (World Bank Group 2022). Nepal is predominantly rural, with more than 80% of the population living in rural areas, and agriculture contributes more than 20% of the GDP. One unique feature of Nepal is its livestock-based economy and the importance of forage as a primary livestock feed. Livestock not only contributes approximately 31% of the agricultural GDP but also utilizes about 11 million hectares of land, including crop lands, for forage (animal feed) purposes (Banshi 2015). Forage is also cultivated as a commercial activity due to the demand–supply gap. Forests also play an important role in supporting human livelihoods and livestock feed. Nepal has been promoting community forest management on a large scale, which has helped improve the livelihoods and forest quality (Ning et al. 2023).

Given the fragile mountainous topography and ecosystems, Nepal is highly vulnerable to climate risks. These risks get accentuated with unplanned settlements and poor infrastructure. Nepal, along with India, is ranked very high in terms of the environmental performance index and environmental health (Huang and Xu 2019). It is the 10th most affected country in the world according to the Climate Risk Index. Given its high dependence on agriculture, 80% of its population is at risk from natural and climate-induced hazards, including extreme heat stress, flooding, and air pollution (World Bank Group 2022). Despite being a water-abundant natural system, it ranks very low in water security among Asian countries. This is likely to exacerbate, as projected temperatures are expected to increase by 0.92–1.07 °C in the medium term (2016–2045) and 1.3–1.8 °C in the long term (2036–2065) compared to the reference period of 1981–2010. At the same time, rainfall is also expected to increase by 2–6% to 8–12% in the medium and long terms, respectively (World Bank Group 2022).

Within South Asia, Nepal is a reasonably well-endowed nation with natural resources, including rainfall, water resources, and forest cover. Average annual rainfall is quite high at 2050 mm and shows a positive anomaly (6 mm) over the period of 30 years (Table 4). The average relative humidity is also high, at above 70%. In line with Asia and other sub-regions, the temperature anomaly is zero, with marginal variations over the years. Nepal has more than 40% of its area under forests, which is higher than the South Asian average. This is reflected in the lower share of agricultural land, which is significantly less than the sub-region's average. This may be the reason for the fragile self-sufficiency in Nepal. Per capita fresh water availability

Table 4 Descriptive statistics of the selected C-N-F indicators for Nepal

Indicator	C-N-F	Mean	SD
AARF in mm	Climate	2,049.8	206.8
AARF Anomaly in mm	Climate	6.2	206.8
AATP Anomaly in C	Climate	0	0.4
RH in %	Climate	71.1	1.5
%FWWD-AGRL	Nature	97.9	0.3
WSTR in %	Nature	8.3	0
% AL	Nature	28.8	0.9
FC in %	Nature	40.8	0.7
FW/CAP in M^3	Nature	7,912.1	939.6
TCS/day/person in Kcal	Food	2,371.1	296.1
SAP in %	Food	1.8	0.2
% MP	Food	16	7.7
GDP/Capita/Year in US$	Economic	661.2	192.3
Nitrogen use in Kgs/ha	Intensification	25.2	16.3
Phosphate use in Kgs/ha	Intensification	10.5	7.5
Potash use in Kgs/ha	Intensification	1.2	1.3
NPK use in Kgs/ha	Intensification	36.9	24.7
Pesticide use in Kgs/ha	Intensification	0.1	0.1

is substantially lower compared to the South Asian average, although water stress is marginal. Nepal uses about 98% of its freshwater for agricultural purposes, which is the highest for any Asian country. Despite the high water use, the intensification of agriculture is much lower in terms of fertilizer and pesticide use per hectare of land. This, coupled with a lower proportion of agricultural land, could be the reason for the moderate calorie supply. The extent of malnutrition is also high at 16% (Table 4).

3.1 C-N-F Nexus: Trends

The trajectories of the key indicators for C-N-F over the 30 year period are examined separately for the climate, nature, and food dimensions. Trends in climate indicators revealed that only rainfall anomalies have shown wide variations over the 30 years, while temperature anomalies and relative humidity have remained relatively stable (Fig. 1). Rainfall anomalies, however, have shown an increasing trend, varying from –463 mm to +455 mm. It is worth noting that rainfall anomalies have shifted only in the positive direction in recent years. Relative humidity has increased marginally from 71.5 to 74.2% over the period of 30 years.

In the case of the nature dimension, most indicators have shown different trajectories (Fig. 2). While per capita water availability has declined sharply, the share of

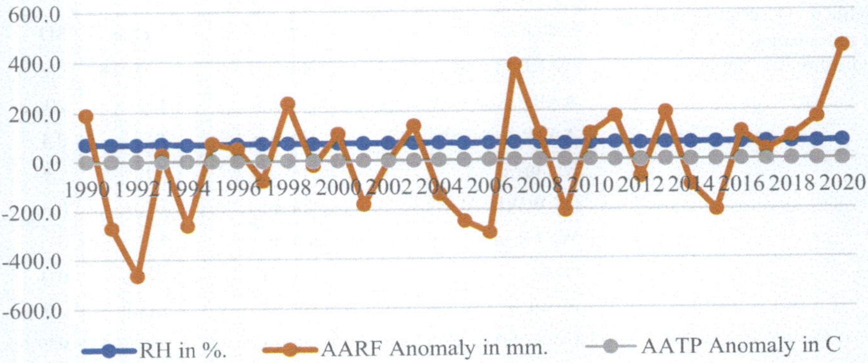

Fig. 1 Trajectories of climate indicators in Nepal. *Source* https://ourworldindata.org/climate-change

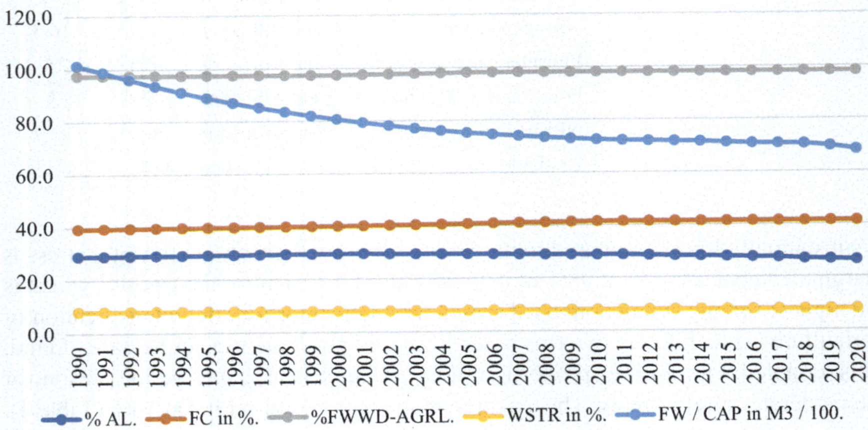

Fig. 2 Trajectories of nature indicators in Nepal. *Source* https://ourworldindata.org/climate-change

agricultural water use has shown an upward trend. This is despite the decline in the share of agricultural land over the three-decade period. This indicates that irrigation intensity has increased over the period. There has been a marginal increase in the forest cover. Water stress has remained relatively stable over the period.

In the case of food indicators, the total calorie supply has increased substantially from approximately 2000–2900 kcal over the past 30 years (Fig. 3). The share of animal protein in total protein has also increased over this period. These improvements are well reflected in the sharp decline in the extent of malnutrition. The average calorie supply over a period of 30 years recorded a sharp increase, despite Nepal

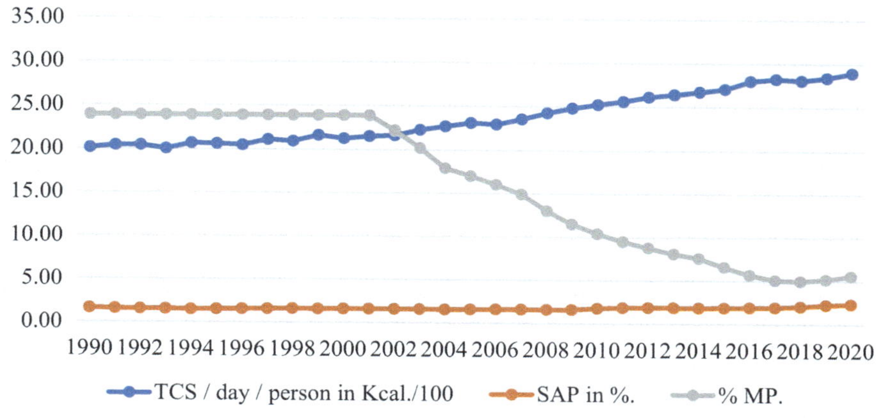

Fig. 3 Trajectories of food indicators in Nepal. *Source* https://ourworldindata.org/climate-change

facing food security issues due to declining crop lands and low intensification (technology penetration). The reason could be due to proactive policies that focus on increasing food availability and food security, irrespective of self-sufficiency.

The trends in the core indicators of the C-N-F dimensions support the C-N-F nexus hypothesis in the context of Nepal. Forest cover has increased along with marginal changes in temperature anomalies (Fig. 4). These changes, taken together, have moved in the same direction as calorie supplies. This supports the C-N-F nexus hypothesis, although not in a systematic manner. The increased calorie supply is not directly associated with improved natural conditions and stable or reduced climate risks (no increase in temperature anomalies). At the same time, Nepal is facing an increase in climate risks due to the rising incidence of floods and other natural disasters. In recent years, climate risks have adversely impacted food security (World Bank Group 2022). Public policy and the policy environment have contributed to ensuring an improved calorie supply in Nepal. These can be seen as indirect linkages that support the C-N-F nexus. This is further probed with the help of the quantitative assessment presented in the following section.

3.2 *C-N-F Nexus: Quantitative Assessment*

Nepal falls on the other extreme in our assessment of the C-N-F nexus. Nepal is a lower-middle-income country with moderate GDP growth, widespread poverty, and high migration. It is food insecure, despite having reasonable natural endowments, due to the lack of technology spread and penetration. In terms of environmental performance, Nepal is also among the highest-ranking countries in the world. It would be interesting to examine the C-N-F linkages in the context of a backward economy. While the trend analysis has already provided some interesting insights,

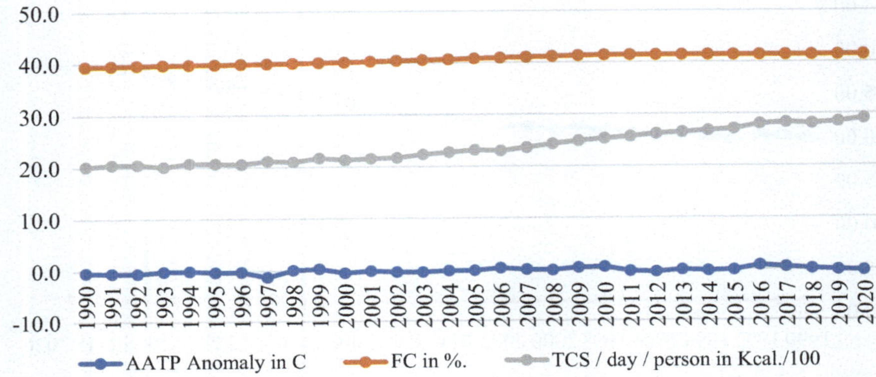

Fig. 4 Climate, nature, and food trajectories in Nepal. *Source* https://ourworldindata.org/climate-change

this section assesses the linkages quantitatively with the help of correlation coefficients. The assessment is carried out between and with each of the C-N-F dimensions, covering four periods: the long-term period of 30 years (1990–2020) and the three sub-periods, which cover the three consecutive decades (1990–2000, 2001–2010, and 2011–2020). The number of indicators that turned out to be significant in all three dimensions. While the detailed correlation matrix is presented in the appendix (Table 9), only the significant correlation coefficients are presented and discussed here (Table 5).

3.3 Climate

Temperature anomalies, representing the climate dimension, are positively associated with GDP, forest cover, agricultural water use, total calorie supply, and pesticide use over the period of 30 years (Table 5). Climate (temperature) and nature (forest), as well as climate and food (total calorie supply), are moving in the same direction. While the positive climate-nature linkage is due to proactive forest policies, the direct relation between climate and food supports (Mendelsohn's 2014) observations in the case of Asia. The positive relationship between temperature and pesticide use suggests that chemical use tends to increase climate risks through its adverse impact on biodiversity. Water availability for agriculture complements chemical use in agriculture, thereby explaining the positive association between the two. On the other hand, the negative association between temperature anomalies and water availability and water stress is in line with expectations. All these indicators are consistent in their signs over different periods. During period three (2001–2010), additional indicators, such as the share of animal protein (SAP), relative humidity, and proportion

Table 5 Significant correlation coefficients of the C-N-F nexus indicators for Nepal

Period	Climate—Temperature	Nature—% Forest Cover	Food—Total Calorie Supply/capita/day (Kcal)
P1 (1990–2020)	TCS/day/person in Kcal:0.39**; % MP.: –0.42**; FC in %: 0.45**; %FWWD-AGRL: 0.44**; WSTR in %.: –0.44**; Pesticide use in Kgs/ha.:0.32***; FW/CAP in M^3.: –0.42***; GDP/Capita/Year in US$:0.34***	TCS/day/person in Kcal: 0.93*; SAP in %.: 0.71*; % MP.: –0.94*; % AL.: –0.65*; AARF Anomaly in mm:0.29***; AATP Anomaly in C:0.45**; %FWWD-AGRL:0.96*; WSTR in %.: –0.95*; Pesticide use in Kgs/ha.:0.82*; Potash use in Kgs/ha.: 0.46**; Phosphate use in Kgs/ha.: 0.28***; NPK use in Kgs/ha: 0.24***; FW/CAP in M^3.: –0.95*; GDP/Capita/Year in US$: 0.88*	SAP in %.: 0.90*; % MP.: –0.98*; % AL.: –0.87*; FC in %:0.0.93*; AARF Anomaly in mm: 0.32***; AATP Anomaly in C:0.39***; %FWWD-AGRL: 0.86*; WSTR in %.: –0.86*; Pesticide use in Kgs/ha.: 0.96*; Nitrogen use in Kgs/ha.: 0.51**; Potash use in Kgs/ha.: 0.73*; Phosphate use in Kgs/ha.: 0.58**; NPK use in Kgs/ha: 0.55**; FW/CAP in M^3.: –0.84*; GDP/Capita/Year in US$: 0.98*
P2 (1990–2000)	%FWWD-AGRL:0.63**; WSTR in %.: –0.59**; Pesticide use in Kgs/ha.: 0.42***; FW/CAP in M3.: –0.77*; GDP/Capita/Year in US$: 0.85*	TCS/day/person in Kcal: 0.86*; RH in %:0.0.46***; % AL.: 1.00*; Pesticide use in Kgs/ha.: 0.76*; Potash use in Kgs/ha.: 0.61**; Phosphate use in Kgs/ha.: 0.44***; NPK use in Kgs/ha: 0.41***; FW/CAP in M^3.: –0.99*; GDP/Capita/Year in US$: 0.99*	% AL.:0.87*; FC in %: 0.0.86*; Pesticide use in Kgs/ha.: 0.69*; Phosphate use in Kgs/ha.: 0.57**; FW/CAP in M3.: –0.84*; GDP/Capita/Year in US$: 0.88*

(continued)

Table 5 (continued)

Period	Climate—Temperature	Nature—% Forest Cover	Food—Total Calorie Supply/capita/day (Kcal)
P3 (2001–2010)	TCS/day/person in Kcal: 0.79*; SAP in %.: 0.70*; % MP.: –0.79*; RH in %:. –0.70*; % AL.: –0.80*; FC in %:0.0.81*; %FWWD-AGRL:0.63**; WSTR in %.: –0.59**; Pesticide use in Kgs/ha.: 0.42***; FW/CAP in M^3.: –0.77*; GDP/Capita/Year in US$: 0.85*	TCS/day/person in Kcal: 0.98*; SAP in %.: 0.65*; % MP.: –0.99*; RH in %:. –0.67*; % AL.:1.00*; AATP Anomaly in C: 0.81*; %FWWD-AGRL: 0.86*; WSTR in %.: –0.84*; Pesticide use in Kgs/ha.: 0.76*; Phosphate use in Kgs/ha.: 0.49***; FW/CAP in M^3.: –0.99*; GDP/Capita/Year in US$: 0.97*	SAP in %.: 0.67; % MP.: –0.98; RH in %: –0.70; % AL.: –0.98; FC in %:0.98; AATP Anomaly in C:0.79; %FWWD-AGRL: 0.81; WSTR in %.: –0.80; Pesticide use in Kgs/ha.: 0.76; Phosphate use in Kgs/ha.: –0.41; FW/CAP in M^3.: –0.96; GDP/Capita/Year in US$: 0.98
P4 (2011–2020)	TCS/day/person in Kcal:0.43***; % MP: –0.56**; Pesticide use in Kgs/ha.: 0.40***	None	SAP in %.: 0.83*; % MP.: –0.95*; RH in %: 0.0.67*; % AL.: –0.95*; AARF Anomaly in mm: 0.43***; AATP Anomaly in C: 0.43***; Pesticide use in Kgs/ha.: 0.92*; Nitrogen use in Kgs/ha.: 0.87*; Potash use in Kgs/ha.: 0.96*; Phosphate use in Kgs/ha.: 0.86*; NPK use in Kgs/ha: 0.89*; FW/CAP in M^3.: –0.91*; GDP/Capita/Year in US$: 0.92*

Note *, **, and *** indicate the level of significance at 1, 5, and 10% levels, respectively
Source Estimated by the authors

of area under agriculture, turned out to be significant. While SAP has shown a positive association with temperature anomalies, it is likely due to the fact that increased consumption of animal protein is detrimental to both nature and the climate. The negative association between the share of agricultural land and temperature anomaly could be attributed to the trade-off between agriculture and forest land use, as forest cover has increased over the period. The recent decade has yielded a smaller number of significant indicators, although with consistent signs.

3.4 *Nature*

Nature (forest cover) is positively associated with food (total calorie supply), the share of animal protein, rainfall anomalies, temperature anomalies, agriculture water use, fertilizer use, pesticide use, and GDP per capita (Table 5). Over the past 30 years, climate factors, especially temperature and rainfall, have complemented forest cover. Forest cover and food (calorie supply) are positively linked, and the share of animal protein is also positively linked to forest cover. This indicates that the forests in Nepal complement the food needs (through forest products and livestock support) and livelihoods (Ning et al. 2023). It also supports the earlier observations that forests support the production of animal protein. Moreover, in the case of Nepal, forests extensively support livestock needs, such as stock fodder (forage). The positive association with chemical use (NPK, pesticides, etc.) is not straightforward. In Nepal, forage is cultivated extensively, including on 30% of the forest area, and is even cultivated on a commercial basis; hence, the use of fertilizers and pesticides tends to be higher, as their use on forage is common (Banshi 2015). The positive association between forest cover and GDP per capita is expected, especially given the policy focus on forest improvement programs in Nepal since the 1990s. On the other hand, forest cover has a negative association with the extent of malnutrition, the share of agricultural land, water availability, and water stress. While the share of agricultural land is expected to move in the opposite direction from forest cover, water availability, and water stress are likely to move in the same direction. Even the extent of malnutrition is inversely linked to forest cover, indicating that biodiversity is not improving the nutrition status. The signs of most indicators have been consistent over the years.

3.5 *Conservation-Focused, Nature-First Economy*

Nepal presents a very interesting case study in terms of geography, demographics, biodiversity, economy, and policy. As a result, the analysis for Nepal yields very interesting and important results, which shed light on how a well-balanced approach to managing the C-N-F nexus is not only possible but also central to facilitating strong, positive outcomes for the land, people, and biomes of a nation. Nepal is home to a rich and diverse flora and fauna, which form the basis of both food systems and other supply chains. This has meant that as the nation and its economy grew, Nepal has managed to prioritize the preservation of its natural resources without undermining the growing needs of its people.

When the correlation analysis is applied to Nepal, the data paint a picture of a nation that is greatly reliant on animal protein to enhance its calorie availability. While this is certainly the case for the rest of Asia as well, the mechanism by which this is achieved lends itself to deeper analysis. Contrary to the rest of Asia, the share of agricultural land is highly negatively correlated to total calorie availability (Nepal: –0.87; Asia: 0.54) for the longer period (1990–2020). This indicates that the growth of agriculture is not necessarily contributing to higher calorie availability in the country. This is supported by data indicating that animal husbandry is a significant contributor to Nepal's food security, suggesting a conflict between the two major land use patterns.

In Nepal, a high share of forest cover correlates with higher calorie availability. While the share of agricultural land grows at the expense of forest cover, the expansion of agriculture in Nepal appears to be substituting grazing lands and fodder lands used for animal husbandry, thus competing with each other. This is further exacerbated by what is predominantly being cultivated, with fodder accounting for a significant share of the country's agricultural output of the country. This establishes the delicate balance maintained between food systems and nature—a strong forest cover can continue to support growing food needs without the need for rapid intensification.

This eco-conservative approach yields obvious climate benefits as well, with temperature and rainfall anomalies generally being low, as well as low correlations with calories and malnutrition (Table 6). Nepal has benefitted from relatively stable climatic conditions, with low disruptive events during the 30 year period, and much weaker correlations across the board when compared to the Asia aggregates. Mendelsohn (2014) notes the potential benefits of further intensification, particularly in the context of rising global temperatures, and highlights this in the case of Asia. However, Nepal may represent an outlier to this pattern, where stable climatic conditions continue to support rising calorie availability. Moreover, the often-criticized challenge of animal protein–rich diets—particularly the carbon and methane emissions associated with large-scale livestock production—can be effectively mitigated through ecological preservation and responsible consumption practices.

It is also important to contextualize the role of water management in this approach to food production, security, and conservation. What is interesting is that water stress data is highly negatively correlated to calorie availability, which is counter to the Asia aggregate. This suggests that while other countries prioritize food production to the extent that the highest production levels are achieved during periods of high water stress and intensification, in Nepal, it is the opposite. Agriculture appears

Table 6 Correlation coefficients between specific C-N-F indicators

Nepal 1990–2020	TCS/day/ person in Kcal	SAP in %	% MP	RH in %	% AL	FC in %	AARF Anomaly in mm
AARF Anomaly in mm	0.32***	0.43**	–0.28***	0.61**	–0.34**	0.29***	–
AATP Anomaly in C	0.39***	0.18	–0.42**	–0.11	–0.23***	0.45**	0.10
%FWWD-AGRL	0.86*	0.62**	–0.91*	–	–	–	–
WSTR in %	–0.86*	–0.62**	0.91*	–	–	–	–

Note *, **, and *** indicate the level of significance at 1, 5, and 10% levels, respectively
Source Estimated by the authors

Table 7 Correlation coefficients between food and intensification practices for Nepal (1990–2020)

Indicator	TCS/day/person in Kcal	SAP in %	% MP
%FWWD-AGRL	0.86*	0.62**	–0.91*
WSTR in %	–0.86*	–0.62**	0.90*
Pesticide use in Kgs/ha	0.96*	0.93*	–0.93*
Nitrogen use in Kgs/ha	0.51**	0.72*	–0.43**
Potash use in Kgs/ha	0.73*	0.90*	–0.64*
Phosphate use in Kgs/ha	0.58**	0.79*	–0.48**
NPK use in Kgs/ha	0.55*	0.77*	–0.46**

Note *, **, and *** indicate the level of significance at 1, 5, and 10% levels, respectively
Source Estimated by the authors

Table 8 Correlation coefficients between share of animal protein and other indicators for Asia (1990–2020)

Indicator	TCS/ day/ person in Kcal	% of Malnourished Population	Nitrogen use in cropland Kgs/ha	Phosphate use in cropland Kgs/ha	Potash use in cropland/ ha	NPK use in cropland Kgs /ha	WSTR in %	Pesticide use cropland Kgs/ha)
SAP in %	0.97*	–0.95*	0.93*	0.02	0.13	0.79*	–0.85*	–0.21***

Note *, **, and *** indicate the level of significance at 1, 5, and 10% levels, respectively
Source Estimated by the authors

to be more reliant on water abundance, and periods of water stress often prioritize human consumption over agricultural production. However, during periods of abundant water availability, sufficient withdrawals for agriculture are accommodated.

As a result, the intensification variables (water stress, pesticide use, and fertilizer use) yield counterintuitive findings (Table 7). This can be attributed mainly to the large part that foraging and open grazing play in the food production systems of Nepal, as well as the application of intensification practices (such as pesticide and fertilizer application) even in forage cultivation. With fodder production accounting for a significant portion of agricultural activity, alongside the usual impacts on calorie availability and malnourished populations, there is a notably higher correlation with the share of animal protein, which is notably lacking in the rest of Asia (Table 8).

4 Conclusions

South Asia, along with Southeast Asia, is the region that has been enjoying the fruits of consistent growth over the past three decades. At the same time, there are wide variations in their growth performance and economic well-being. While

India and Bangladesh are at the forefront, Sri Lanka, Nepal, and other countries are lagging behind. These differences could be attributed to their resource endowments coupled with their policy environment. India and Bangladesh have embarked on a high-growth path, utilizing their service sectors and manufacturing sectors, respectively. Both countries had to adopt intensification paths on their farms to meet their increasing food demand. This has resulted in the severe degradation of natural resources, including land, water, and biodiversity. The proactive forest and resource policies have not been effective in significantly improving natural capital. Although Sri Lanka had initiated economic reforms and advanced in agricultural technologies much earlier, it suffered from prolonged political instability and lost its early advantage in the process. Nepal, however, has been moving on a medium growth path, as the giant economies of India and China overshadow it.

The assessment of the C-N-F nexus at the regional and country levels clearly indicates its presence over a longer period, as well as in recent decades. At the regional level, most countries are moving on the margins of achieving food security and self-sufficiency, although food availability/supply is keeping pace with demand. This is associated with deteriorating natural conditions or natural capital, especially in terms of water resources. Loss of biodiversity is also evident. On the other hand, Nepal has adopted a less intensive path for meeting its food demands. In the absence of crop intensification, forage development for livestock was given importance along with protecting its forest resources and biodiversity. Livestock and forests supported the food needs in Nepal. This suggests that Nepal has managed the C-N-F nexus in a more balanced manner, which may be more due to a lack of choice rather than a deliberate strategy. Nevertheless, this is a more effective way of managing the C-N-F nexus compared to the approaches of India and Bangladesh. Nepal provides some important insights into managing the nexus.

Appendix

See Table 9.

Table 9 Correlation coefficients of the C-N-F nexus indicators for Nepal

Indicators	1990–2020			1990–2000			2000–2010			2011–2020		
	TCS/day/person in Kcal	FC in %	AATP Anomaly in C	TCS/day/person in Kcal	FC in %	AATP Anomaly in C	TCS/day/person in Kcal	FC in %	AATP Anomaly in C	TCS/day/person in Kcal	FC in %	AATP Anomaly in C
TCS/day/person in Kcal	1.00	0.93	0.39	1.00	0.86	0.13	1.00	0.98	0.79	1	–3E–15	0.43
SAP in %	0.90	0.71	0.18	0.01	–0.08	–0.05	0.67	0.65	0.70	0.83	1E–16	–0.08
% MP	–0.98	–0.94	–0.42	0.00	0.00	0.00	–0.98	–0.99	–0.79	–0.95	4E–16	–0.56
RH in %	0.20	0.13	–0.11	0.14	0.46	–0.12	–0.70	–0.67	–0.70	0.67	9E–15	0.21
% AL	–0.87	–0.65	–0.23	0.87	1.00	0.21	–0.98	–1.00	–0.80	–0.95	–3E–15	–0.18
FC in %	0.93	1.00	0.45	0.86	1.00	0.18	0.98	1.00	0.81	0.00	1E + 00	0.00
AARF Anomaly in mm	0.32	0.29	0.10	0.11	0.39	0.28	0.19	0.21	–0.17	0.43	3E–17	–0.02
AATP Anomaly in C	0.39	0.45	1.00	0.13	0.18	1.00	0.79	0.81	1.00	0.43	0E + 00	1.00
%FWWD-AGRL	0.86	0.96	0.44	0.00	0.00	0.63	0.81	0.86	0.63	–	–	–
WSTR in %	–0.86	–0.95	–0.44	0.00	0.00	–0.59	–0.80	–0.84	–0.59	–	–	–
Pesticide use in Kgs/ha	0.96	0.82	0.32	0.69	0.76	0.42	0.76	0.76	0.42	0.92	9E–17	0.40
Nitrogen use in Kgs/ha	0.51	0.20	0.02	0.21	0.34	0.39	0.10	0.02	0.39	0.87	6E–17	0.28
Potash use in Kgs/ha	0.73	0.46	0.13	0.37	0.61	–0.03	–0.10	–0.18	–0.03	0.96	0E + 00	0.21

(continued)

Table 9 (continued)

Indicators	1990–2020			1990–2000			2000–2010			2011–2020		
	TCS/day/person in Kcal	FC in %	AATP Anomaly in C	TCS/day/person in Kcal	FC in %	AATP Anomaly in C	TCS/day/person in Kcal	FC in %	AATP Anomaly in C	TCS/day/person in Kcal	FC in %	AATP Anomaly in C
Phosphate use in Kgs/ha	0.58	0.28	0.05	0.57	0.44	–0.08	–0.41	–0.49	–0.08	0.86	2E–16	0.21
NPK use in Kgs/ha	0.55	0.24	0.04	0.36	0.41	0.26	–0.06	–0.14	0.26	0.89	–4E–16	0.26
FW/CAP in M3	–0.84	–0.95	–0.42	–0.84	–0.99	–0.77	–0.96	–0.99	–0.77	–0.91	5E–15	–0.08
GDP/Capita/Year in US$	0.98	0.88	0.34	0.88	0.99	0.85	0.98	0.97	0.85	0.92	–8E–16	0.19

Note *; ** and *** indicate the level of significance at 1, 5 and 10% levels, respectively
Source Estimated by the authors

References

Banshi S (2015) Present Status and future strategy of forage development in Nepal. J Agric Environ 16. https://doi.org/10.5555/20153323644

Huang B, Xu Y (2019) Environmental performance in Asia: overview, drivers, and policy implications; ADBI Working Paper 990. Tokyo: Asian Development Bank Institute. https://www.adb.org/publications/environmental-performance-asia-overview-drivers-policy-implications

Knerr B (2017) Nepal: a short country profile. In: Knerr B (ed) International labor migration and livelihood security in Nepal: considering the household level. Kassel University Press GMBH. https://books.google.co.in/books?hl=en&lr=&id=xDIuDwAAQBAJ&oi=fnd&pg=PA9&dq=Socioeconomic++and+environmental+profile+of+Nepal&ots=s4NvdJKCeE&sig=U7k5SvCzp4sBoJ4yKMRhfmTrS4E#v=onepage&q&f=false

Lal DG (n.d.) Population and Environmental Issues in South Asia; General Survey, https://www.researchgate.net/profile/Gouranga-Dasvarma/publication/283487526_Population_and_Environmental_Issues_in_South_Asia/links/563a0cb808ae337ef29827c7/Population-and-Environmental-Issues-in-South-Asia.pdf

Mendelsohn R (2014) The impact of climate change on agriculture in Asia. J Integr Agric 13(4):660–665. https://doi.org/10.1016/S2095-3119(13)60701-7

Ning C, Subedi R, Hao L (2023) Land use/cover change, fragmentation, and driving factors in Nepal in the Last 25 Years. Sustainability 15:6957. https://doi.org/10.3390/su15086957

Reddy VR, Rahut DB, Sonobe T (2025) Future of small holder farms in Asia. Asian Development Bank Institute, Tokyo, Japan. https://www.adb.org/sites/default/files/publication/1041036/future-smallholder-farming-asia.pdf

World Bank Group (2022) Nepal Country Climate and Development Report. CCDR Series. World Bank Group. http://hdl.handle.net/10986/38012. License: CC BY-NC-ND

Chapter 7
South East Asia Case Study: Vietnam

1 Introduction

Southeast Asia, comprising nine countries (Brunei, Cambodia, Indonesia, Laos, Malaysia, Myanmar, the Philippines, Thailand, and Vietnam), is one of the economically prosperous regions. While these nations are smaller than the countries in the East and South Asian sub-regions, they make significant contributions to the biological makeup of the Asian averages. One interesting aspect of the region is its homogeneity among countries in terms of climate, plant and animal life, as well as socioeconomic progress. Southeast Asia is among the major rice-producing regions, although agriculture contributes marginally to the GDP. A unique economic feature of this region is its dependence on tourism and the financial sectors. Together, they account for a major share of the region's GDP. Thus, Southeast Asia's booming tourism and financial services, which contribute to its economic growth, provide an interesting ground for assessing the C-N-F nexus. To understand the region, it is essential to recognize that it is managing tourism-driven economic development and environmental sustainability. It is observed that the impact of tourism growth on environmental factors is evident in terms of renewable energy use, carbon dioxide (CO_2) emissions, and ecological footprint (Ha Van 2024). As a result, the region is being affected by climate change, which is impacting agriculture through frequent droughts and floods. Agriculture, in turn, also contributes to GHG emissions through unsustainable farm practices, such as inefficient water use and stubble burning.[1]

As the region lies between the tropics, and the nature of the Oceans (shallow and warm) makes it an ideal environment for rich coastal biodiversity, including fish, corals, seaweeds, and other marine products. This region is also less affected by coastal storms, hurricanes, and typhoons.[2] Despite the rich biodiversity in the region, it is ecologically vulnerable. Most countries in the region are ranked high in

[1] https://www.irri.org/where-we-work/countries/southeast-asia.

[2] https://asiasociety.org/education/introduction-southeast-asia.

V. R. Reddy et al., *Strengthening the Climate–Nature–Food Nexus*,
https://doi.org/10.1007/978-3-032-29423-4_7

terms of the environmental performance index and environmental health (Huang and Xu 2019). Thus, Southeast Asia presents an interesting case for assessing the C-N-F nexus in the context of service-driven developing economies.

2 Quantitative Assessment of C-N-F Nexus: Southeast Asia

Southeast Asia has the highest average annual rainfall among all the sub-regions of Asia, although rainfall anomalies have recorded an average decline of 3.1 mm over the past three decades (Table 1). It has high average humidity levels of above 70%. Southeast Asia is the only sub-region that has reported a marginal decline in the average temperature anomaly. Southeast Asia has the highest percentage of forest cover (52.7%) and the second-highest water availability per capita among the sub-regions. The region has a relatively low proportion (24.4%) of area under agriculture. This could be due to urbanization in the region—it has more than 50% of its population living in urban areas, and it is one of the fastest urbanizing regions in Asia. Given the good rainfall, forest cover, and water availability, the region has very low water stress levels (10%). The proportion of total water used for agriculture is as high as 73%, despite the fact that only a quarter of the land is allocated to agriculture. Crop intensification is also reflected in the use of chemical fertilizers and pesticides per unit of land. Southeast Asia has a reasonably high per capita GDP and a good calorie supply. However, the extent of malnutrition is also high, at 15%, indicating low nutritional levels in the available food. In terms of the incidence of poverty, there are wide variations among countries in the region, although most countries are reporting declining poverty rates in recent years. Poverty reduction continues to be a policy challenge in countries like Cambodia in the region. Thus, Southeast Asia is a fast-growing, developing economy with good and diverse natural endowments. To assess the C-N-F nexus linkages in the region, a quantitative analysis is conducted using regional-level correlation coefficients of various indicators associated with C-N-F dimensions. The number of indicators representing the three dimensions has turned out to be significant in all four periods, although the direction and strength of the linkages vary over the periods (Table 2). The region yields some interesting relationships. Indicators representing each of the C-N-F dimensions and their interlinkages across the four periods are discussed below.

2.1 *Climate*

Food indicators, such as total calories, total protein supply, and the share of animal protein, have demonstrated a positive correlation with temperature anomalies over the period of 30 years (Table 2). The indicators directly linked to food production, namely the percentage area under agriculture and NPK use, have shown a positive association

Table 1 Descriptive statistics of selected indicators for south East Asia

Indicator	C-N-F	Descriptive statistics	
		Mean	SD
AARF in mm	Climate	2377	187.8
AARF anomaly in mm	Climate	−3.1	187.8
AATP anomaly in C	Climate	−0.01	0.34
RH in %	Climate	79.65	0.78
%FWWD-AGRL	Nature	73.08	8.3
WSTR in %	Nature	10.48	1.9
% AL	Nature	24.4	2.1
FC in %	Nature	52.7	1.5
FW/CAP in M^3	Nature	14,963	2367
TCS/day/person in Kcal	Food	2550	204.1
TPS/capita/day in grams	Food	66.4	8.9
APS/capita/day in grams	Food	25.2	5
SAP in %	Food	3.81	0.5
% MP	Food	15	6.4
GDP/capita/year in US$	Economic	6353.5	319.1
Nitrogen use in Kgs/ha	Intensification	46.11	7.3
Phosphate use in Kgs/ha	Intensification	19	3.9
Potash use in Kgs/ha	Intensification	24.66	6.2
NPK use in Kgs/ha	Intensification	89.7	16.3
Pesticide use in Kgs/ha	Intensification	4.4	0.8

with temperature anomalies. These relationships are on the expected lines, as intensive farm practices and an increased supply of animal protein are likely to adversely impact climate or temperature anomalies. Similarly, water stress and temperature anomalies are moving in the same direction. On the other hand, nature indicators like forest cover and water availability have shown a negative relation with temperature anomalies, as expected. Only relative humidity and extent of malnutrition move in the opposite direction to temperature anomalies (Table 2). While the negative association between temperature anomalies and relative humidity is expected, the negative association of the extent of malnutrition could be due to changes in the microenvironment, such as hygiene conditions. While sub-periods two (1990–2000) and three (2001–2010) have revealed that only a few indicators became significant, and most of them showed consistent signs. The only exception is the GDP, which turns up with a negative sign. During the recent decade, some additional indicators, such as pesticide use and rainfall anomalies, have become significant with expected signs.

Table 2 Significant correlation coefficients of the C-N-F nexus indicators for the South East Asia sub-region

Period	Climate—temperature	Nature—% forest cover	Food—total calorie supply/capita/day (Kcal)
P1 (1990–2020)	FW/CAP in M^3: –0.70; FC in %.: –0.75; NPK use in Kgs/ha: 0.58; WSTR in %.: 0.66; % AL: 0.74; GDP/capita/year in US\$.: 0.62; TCS/day/person in Kcal: 0.74; % MP: –0.71; RH in %: –0.34; TPS in grams/capita: 0.74; SAP in %: 0.69	FW/CAP in M^3: 0.98; NPK use in Kgs/ha: –0.83; WSTR in %.: –0.91; % AL: –0.98; GDP/capita/year in US\$.: –0.79; SAP in %.: –0.95: TCS/day/person in Kcal: –0.98; MP in %: 0.93; AATP anomaly: –0.75	AATP anomaly in C: 0.74; FC in %.: –0.98; FW/CAP in M^3: –0.95; NPK use in Kgs/ha: 0.76; WSTR in %.: 0.96; % AL: 0.99; MP in %: –0.98; GDP/capita/year in US\$.: 0.85; SAP in %: 0.98
P2 (1990–2000)	GDP/capita/year in US\$.: –0.43; RH in %: –0.57	FW/CAP in M^3: 1.00; NPK use in Kgs/ha: –0.98; Pesticide: –0.70; GDP/capita/year in US\$.: 0.66; SAP in %.: 0.82; % AL: –0.93; TCS/day/person in Kcal: –0.92; % MP: –0.50; AARF in mm: –0.40; RH in %: –0.50; %FWWD-AGRL: 0.92; TPS in grams/capita: –0.95	AARF anomaly in mm.: 0.62; RH in %:0.52; FC in %.: –0.92; FW/CAP in M^3: –0.93; NPK use in Kgs/ha:0.89; Pesticide use: 0.71; % AL: 0.81; %FWWD-AGRL: –0.86; GDP/capita/year in US\$.: –0.70; SAP: 0.88; AARF in mm: 0.62
P3 (2001–2010)	Nitrogen use in Kgs/ha use: 0.50; TPS in grams/capita: 0.48; SAP in %: 0.41; RH in %: –0.40	FW/CAP in M^3: 1.00; NPK use in Kgs/ha: –0.47; WSTR in %.:–0.89; % AL: 1.00; SAP in %.: –0.97; TCS/day/person in Kcal: –0.99; MP in %:1.00; AARF in mm: –0.62; AARF anomaly: = 0.62; %FWWD-agrl: –0.92	AARF Anomaly in mm.: 0.56; FC in %.: –0.99; FW/CAP in M^3: –0.99; NPK use in Kgs/ha: 0.53; WSTR in %.: 0.86; % AL: 0.99.; AARF in mm: 0.56; %FWWD-AGRL: –0.88; % MP: –0.99; GDP/capita/year in US\$.:0.46; SAP in %: 0.97
P4 (2011–2020)	FW/CAP in M^3: –0.78; FC in %.: –0.80; WSTR in %.: 0.71; % AL: 0.81; Pesticide use in Kgs/ha use: 0.64. GDP/capita/year in US\$.: 0.67; TCS/day/person in Kcal: 0.79; % MP: –0.80; AARF Anomaly in mm.: –0.61; RH in %: –0.69; APS in grms/capita: 0.64	FW/CAP in M^3: 1.00; NPK use in Kgs/ha: –0.45; WSTR in %: –0.91; Pesticide use in Kgs/ha use: –0.56; % AL: –0.98; GDP/capita/year in US\$.: –0.86; Animal protein: –0.37; –0.99; MP in %:0.99; RH in %: 0.45; AATP anomaly: –0.80; APS in grams/capita: –0.76	AATP anomaly in C: 0.79; RH in %: –0.43; FC in %.: –0.99; FW/CAP in M^3: –1.00; NPK use in Kgs/ha: 0.42; WSTR in %.: 0.89; Pesticide use in Kgs/ha use: 0.58; % AL: 0.98; % MP: –0.98; GDP/capita/year in US\$.: 0.88; TPS in grams/capita: 0.92; APS in grams/capita: 0.72

Note *, **, and *** indicate the level of significance at 1%, 5%, and 10% levels, respectively
Source Estimated by the authors

2.2 Nature

Only two indicators, namely water availability and the extent of malnutrition, turned out to be significant with a positive sign in relation to forest cover (Table 2). It is true that forest cover complements water availability as part of its ecosystem services. The positive linkage with the extent of malnutrition indicates that forests support the supply of nutritious food. All other indicators showed a negative association with forest cover that is in the right direction. The negative sign of the share of agriculture land reflects the trade-off between forest and agricultural land. The use of NPK has shown a negative association with forest cover due to its adverse impacts on biodiversity. Water stress and forest cover move in opposite directions due to the declining ecosystem services associated with forest cover. Increased food availability means the expansion of agricultural lands, which comes at the cost of forest cover, and hence the negative linkage between these two. Increased animal protein often comes at the cost of forest cover, i.e., degradation due to forage. The average temperature anomaly appears to have a negative association with forest cover over the longer period. And there is a negative association with GDP, as they are often expected to move in the opposite direction, unless public policy intervenes. All these linkages are consistent across the sub-periods.

2.3 Food

In the case of food (total calorie supply), important indicators such as the share of agricultural land, NPK use, GDP per capita, share of animal protein, temperature anomaly, and water stress have a positive association. Of these, only the positive association between water stress and food availability suggests that crop production systems, particularly rice, appear to be performing well under water-stress conditions. The relationships are consistent over the sub-periods. Temperature/rainfall anomalies have turned out to be positive in all periods. Forest cover has shown a negative association with total calorie supplies, vindicating the trade-off between forest and agricultural land. The positive association between temperature anomalies and total calorie supply is similar to that observed at the aggregate level (Asia). And the negative relationship between total calorie supply and the extent of malnutrition, which supports the earlier observations in the case of Asia. It is worth noting that GDP per capita, which turned out to be significant and consistent with a positive sign, is the case in most Asian countries. The share of animal protein and total protein supply contributes substantially to food availability in Southeast Asia. In the following section, a comparative assessment of Asia and Southeast Asia is discussed.

2.4 Service Sector Driven Developing Economies and C-N-F Nexus

Southeast Asia is home to some of the most biodiverse and fertile biomes. As such, East Asia and Southeast Asia contribute significantly to the planet in terms of environmental services. Given its temperate climate, it is also home to abundant rainfall, verdant forests, and a rich array of flora and fauna, including marine life. As a result, the data tends to precipitate the cost–benefit relationships between the C-N-F nexus. The Southeast Asia experience suggests that, in a region of abundance, growing needs may start to encroach on the established balance between the three (C-N-F) factors. And the disturbed balance could start to degrade the various positive feedback loops, eventually resulting in a net negative effect of the activity on the C-N-F.

Southeast Asian nations have a highly varied mix of approaches to managing developmental challenges, including food security, infrastructure development, ecological conservation, and economic growth. These circumstances may help explain some of the intricate correlation findings for this sub-region. The region's highly moist baseline climate results in a stronger correlation between relative humidity and average annual rainfall anomalies, as well as temperature anomalies (Table 3). The baseline humid nature of the prevailing climatic conditions also means that the relative humidity and temperature anomaly have a weaker correlation for Southeast Asia.

While the correlations for temperature anomaly to calories (0.73) are along the same lines as the rest of Asia and the other sub-regions, the forest cover to calorie availability correlation goes strongly against the remaining sub-regions (all of them have positive correlations of varying strengths) though it is in line with the Asia aggregate (which has a relatively weak negative correlation of −0.42) (Table 4). With a highly negative correlation between forest cover and calorie availability of −0.98, this region clearly highlights the trade-offs that must be made to improve caloric supply by sacrificing forest cover to reclaim agricultural land. This is corroborated by the extremely high correlation between the area under agriculture and the calorie availability. These correlations, however, have become more pronounced in the last

Table 3 Correlation coefficients between relative humidity and other nature indicators in Asia and Southeast Asia (1990–2020)

Indicator	Asia			Southeast Asia		
	AARF in mm	AARF anomaly in mm	AATP anomaly in C	AARF in mm	AARF anomaly in mm	AATP anomaly in C
RH in %	0.09	0.07	−0.63*	0.09	0.86*	−0.34***

Note *, **, and *** indicate the level of significance at 1%, 5%, and 10% levels, respectively
Source Estimated by the authors

Table 4 Correlation coefficients between total calorie supply and other nature indicators in Asia and Southeast Asia for periods 1990–2020 and 2011–2020

	Asia		Southeast Asia	
	1990–2020	2011–2020	1990–2020	2011–2020
FC in %	−0.42**	−0.99*	−0.98*	−0.99*
% AL	0.54**	0.96*	0.99*	0.98*
Nitrogen use in Kgs/ha	0.93*	0.92*	0.66*	0.45**
Phosphate use in Kgs/ha	-0.09	0.67*	0.77*	0.49**
Potash use in Kgs/ha	0.08	0.59**	0.73*	0.16
NPK use in Kgs/ha	0.75*	0.92*	0.76*	0.42**
WSTR in %	−0.89*	0.15	0.96*	0.89*
Pesticide use in Kgs/ha	−0.14	0.69*	0.09	0.58**

Note *, **, and *** indicate the level of significance at 1%, 5%, and 10% levels, respectively
Source Estimated by the authors

decade, as needs continued to grow, followed by the widespread adoption of intensification practices in agriculture, as evidenced by the variables of intensification (Table 4).

The correlation coefficients reveal a clear picture in Southeast Asia. Although the early gains from intensification were substantial, over time, the value primarily came from water withdrawals, as the economies began to accommodate high-water-consuming industries, such as textiles and garments. However, in Asia, while early gains were limited, the last decade saw the accumulation of gains from continued intensification (in some areas). This is attributed to the scale application of intensification technologies and water infrastructure. For instance, the weaker correlation, particularly with water stress levels, suggests that increased calorie availability was not achieved despite high water stress levels. This is in stark contrast to the trends for Southeast Asia.

The remaining correlation values largely align with the trends for the other Asian sub-regions and the Asia aggregates. It is encouraging that the broad trends and interactions across the regions are consistent, establishing that the nature of the interactions between the factors is consistent, despite varying prevailing circumstances. The differences in the outcomes are largely a result of these circumstances, and more particularly, how they were approached and handled by policymakers and governance.

3 C-N-F Nexus: The Case of Vietnam

Vietnam is one of the fastest-growing nations in the region. Its consistent export-led growth has transformed Vietnam from a lower-income economy to a lower-middle-income economy. It has shifted from an agriculture-based to a manufacturing-driven economy. The contribution of agriculture is only 12% of the GDP, although agriculture continues to be critical as a safety net, especially during crises (United Nations 2024). The consistent economic performance helped reduce poverty in recent years, as the poverty rate was 3.4% in 2023.[3] Such a transformation has come from the degradation of the natural systems. Despite being the 16th most biodiverse country in the world, Vietnam faces significant biodiversity loss and ecosystem degradation. Vietnam is ranked very high in terms of the environmental performance index, environmental health, and ecosystem vitality (Huang and Xu 2019). Deforestation, habitat fragmentation, unsustainable land conversion, pollution, illegal wildlife trade, and climate change are prime drivers of degradation. As a result, Vietnam faces critical climate challenges, including biodiversity loss, natural resource depletion, water scarcity, and serious air pollution in large cities. It is ranked among the top 20 greenhouse gas (GHG) emitting countries as its rapid growing economy continues to be powered by fossil fuels (United Nations 2024). It is observed that Vietnam's economic growth and nature of agricultural development have led to higher utilization of energy and emissions, resulting in increased climate risks (Raihan et al. 2025).

Over the 30-year period, the average annual rainfall is approximately 2031 mm, with significant variations (Table 5). Average annual rainfall anomalies have been on the rise over the period, while temperature anomalies remain unchanged, which is consistent with most sub-regions and case study countries. The relative humidity is very high, at approximately 80%. Forest cover in Vietnam accounts for 40% of the total area, which is significantly below the sub-regional average of above 50%. Per capita water availability in Vietnam is significantly below the sub-regional average. Water stress is also higher in Vietnam. Given the lower forest cover, the proportion of the area allocated to agriculture is higher in Vietnam, at 30.6%. The share of agriculture water use is 94%, which is significantly higher than the sub-regions' average. High water use in agriculture is complemented by high intensity of NPK and pesticide use. Despite the high intensification of food availability in Vietnam, it is lower compared to the sub-region average. Similarly, the average GDP per capita is substantially lower in Vietnam (Table 5). These long-term averages do not accurately reflect recent changes or trends; therefore, the following section presents a visualization of the trajectories of C-N-F core indicators.

3.1 C-N-F Nexus: Trends

As in the case of other countries and sub-regions, only rainfall anomalies have shown wide variations over the period of 30 years, while temperature anomalies and relative humidity have remained stable, though temperature anomalies have indicated a marginal increase (Fig. 1). Whereas rainfall anomalies have shown an increasing

[3] https://www.adb.org/where-we-work/viet-nam/poverty.

Table 5 Descriptive statistics of the selected C-N-F indicators for Vietnam

Indicator	C-N-F	Mean	SD
AARF in mm	Climate	2,031.1	153.5
AARF anomaly in mm	Climate	4.8	153.5
AATP anomaly in C	Climate	0.0	0.4
RH in %	Climate	79.6	0.9
%FWWD-AGRL	Nature	94.0	1.1
WSTR in %	Nature	17.2	1.1
% AL	Nature	30.6	6.3
FC in %	Nature	40.0	5.1
FW/CAP in M^3	Nature	4,436.8	529.5
TCS/day/person in Kcal	Food	2,492.0	346
SAP in %	Food	3.5	1.1
% MP	Food	14.1	5.3
GDP/capita/year in US$	Economic	1,729.5	789.2
Nitrogen use in Kgs/ha	Intensification	125..4	25.7
Phosphate use in Kgs/ha	Intensification	62.4	20.6
Potash use in Kgs/ha	Intensification	37.3	18.6
NPK use in Kgs/ha	Intensification	225.1	54.6
Pesticide use in Kgs/ha	Intensification	3	0.8

trend that varied from −367 mm to +349 mm. It is worth noting that rainfall anomalies have shifted in a positive direction in recent years. Relative humidity has declined only marginally over the past 30 years.

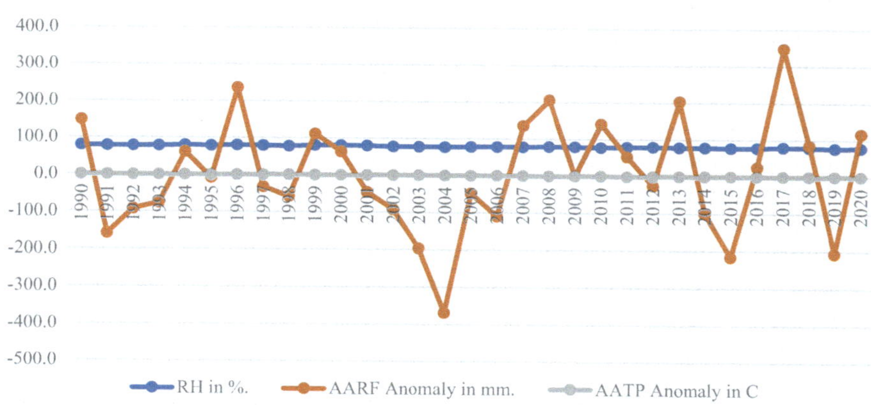

Fig. 1 Trajectories of climate indicators in Vietnam. *Source* https://ourworldindata.org/climate-change

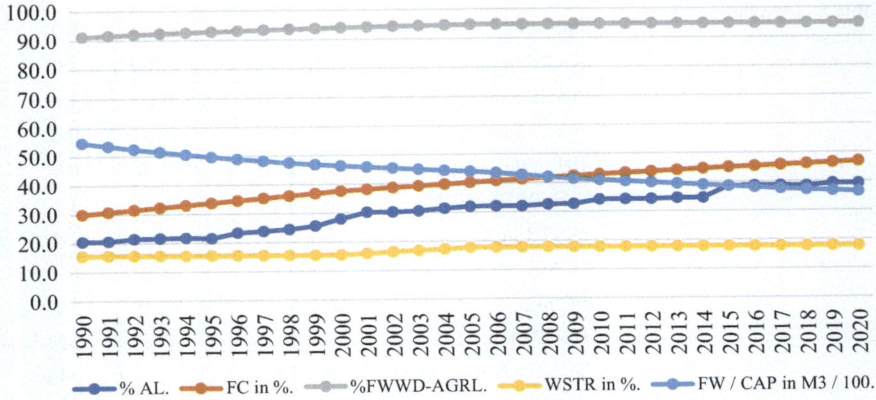

Fig. 2 Trajectories of nature indicators in Vietnam. *Source* https://ourworldindata.org/climate-change

In the case of the nature dimension, all the indicators, except one, have shown increasing trends (Fig. 2). Forest cover has increased substantially from 30% in 1990 to 47% in 2020. While the per capita water availability has declined sharply, the share of agricultural water use has shown an upward trend over the past 30 years. This could be attributed to the increasing trend in the share of area under agriculture. This is reflected in the increased water stress, though marginally over the period.

In the case of food indicators, the total calorie supply has increased substantially from approximately 2000 to 3000 kcal over the past 30 years (Fig. 3). The share of animal protein in total protein has increased by almost three times over the period. These improvements are well reflected in the sharp decline in the extent of malnutrition. The increased average calorie supply over the 30-year period is primarily due to the expansion of agricultural areas, coupled with higher farm intensification. This is despite the fact that the contribution of agriculture to Vietnam's GDP has been declining over the years. The reason could be due to the intensification policies that focus on increasing crop productivity (Fig. 4).

The trends in the core indicators of the C-N-F dimensions support the nexus hypothesis in the context of Vietnam. The sharp rise in forest cover appears to have contributed to climate stability, i.e., marginal changes in temperature anomalies. Together, these changes have led to increased calorie supplies. This endorses the C-N-F nexus hypothesis, though the linkages are not straightforward. It may be argued that the improved natural capital and biodiversity (forest cover) are helping Vietnam address climate risks effectively and enhance calorie supplies. This is achieved through a higher allocation of water to agriculture, coupled with greater intensification, despite declining water availability per capita. For the improved biodiversity (forest cover) to have a direct impact on water availability. These intricate linkages are further probed with the help of quantitative assessment in the next section.

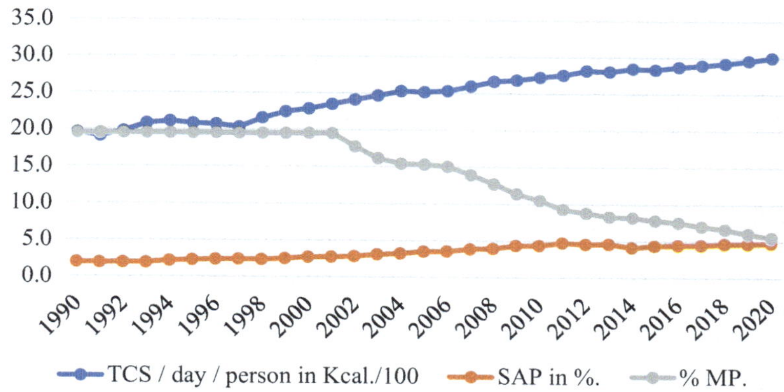

Fig. 3 Trajectories of food indicators in Vietnam. *Source* https://ourworldindata.org/climate-change

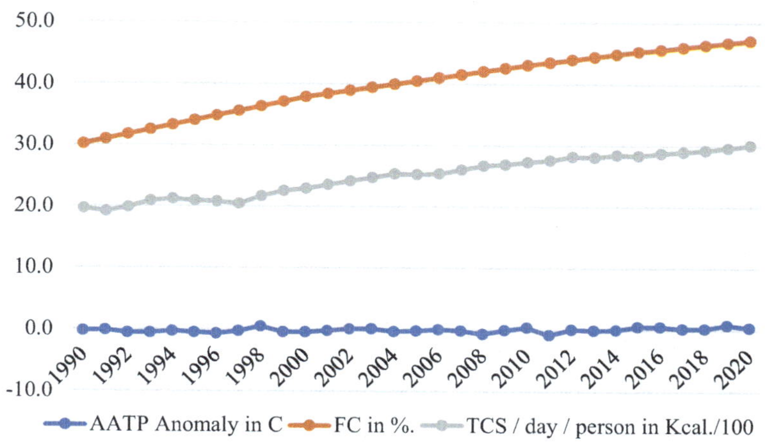

Fig. 4 Climate, nature, and food trajectories in Vietnam. *Source* (https://ourworldindata.org/climate-change)

3.2 C-N-F Nexus: Quantitative Assessment

Vietnam, like Nepal, is a lower-middle-income country. However, it differs in terms of agricultural development (highly intensive and productive) and the structure of the economy, with a higher share of non-agricultural sectors in GDP, and a marginal contribution from agriculture. Vietnam, despite being a lower-middle-income country, is consistently growing with a very low incidence of poverty. Using its natural endowments judiciously and effectively, it is relatively more food secure due to the technology penetration. At the same time, Vietnam is not managing its environment effectively, despite having good natural endowments. It is also observed that its growth path is putting pressure on resources and degrading the environment, making it unsustainable (Raihan et al. 2025). At the same time, a substantial increase in forest cover reflects pro-environmental policies. It would be interesting to examine the C-N-F linkages in the context of a progressive but resource-intensive economy. While the trend analysis has already provided some interesting insights, this section assesses the linkages quantitatively with the help of correlation coefficients. The assessment is carried out across and within each of the C-N-F dimensions, covering four periods: the long-term period of 30 years (1990–2020) and the three sub-periods, which span three consecutive decades (1990–2000, 2001–2010, and 2011–2020). It is worth noting that none of the indicators in all three dimensions proved significant in period three (2000–2010). While the detailed correlation matrix is presented in the appendix (Table 8), this section presents only the significant correlation coefficients and discusses them (Table 5).

3.3 Climate

In the case of the linkages between climate (temperature anomalies) and the remaining indicators, none of the indicators turned significant in periods two and three. Over the 30 years, temperature anomalies have a positive association with total calorie supply, the share of animal protein, the percentage of agricultural land, forest cover, agricultural water use, water stress, NPK use, and GDP per capita (Table 5). Climate (temperature) and nature (forest), as well as climate and food (total calorie supply), are moving in the same direction. While the direct relationship between climate and food supports Mendelsohn's (2014) observations in the case of Asia, the positive climate-nature linkage is due to the proactive policy environment towards afforestation. The positive relationship between temperature and NPK use is due to the fact that chemical use tends to increase climate risks through its adverse impact on biodiversity. Since a higher proportion of agricultural land requires higher chemical inputs, there is a positive association. Water use for agriculture complements chemical use in agriculture, and hence, there is a positive association between these two. On the other hand, temperature anomalies and water stress are moving in the same direction, as increased temperatures not only increase water demand but also reduce water supplies. The share of animal protein in total calories has shown a positive association with temperature anomalies, likely because increased consumption of animal protein is detrimental to nature and thus increases climate risks. The extent of malnutrition, water availability per capita, and relative humidity have shown a negative association with temperature anomalies, which is expected. All these indicators are consistent in their signs in period three (2011–2020) (Table 6).

Table 6 Significant correlation coefficients of the C-N-F nexus indicators for Vietnam

Period	Climate—temperature	Nature—% forest cover	Food—total calorie supply/capita/day (Kcal)
P1 (1990–2020)	TCS/day/person in Kcal.: 0.57**; SAP in %.:0.51**; % MP.: – 0.59**; RH in %.: – 0.41**; % AL.: 0.62**; FC in %.: 0.58**; %FWWD-AGRL: 0.43**; WSTR in %.: 0.46**; Nitrogen use in Kgs/ha.: 0.40***; Potash use in Kgs/ha.: 0.40***; NPK use in kgs/ha.: 0.42**; FW/CAP in M3.: −0.58**; GDP/capita/year in US$.:0.65*	TCS/day/person in Kcal.: 0.99*; SAP in %.: 0.97*; % MP.: −0.93*; % AL.: 0.99*; AATP Anomaly in C.: 0.58**; %FWWD-AGRL: 0.92*; WSTR in %.: 0.90*; Nitrogen use in Kgs/ha.: 0.39***; Potash use in Kgs/ha.: 0.88*; Phosphate use in Kgs/ha.: 0.78*; NPK use in kgs/ha.: 0.78*; FW/CAP in M3.: −1.00*; GDP/capita/year in US$.: 0.95*	SAP in %.:0.98*; % MP.: −0.95*; % AL.: 0.98*; FC in %.: 0.99*; AATP Anomaly in C.: 0.57*; %FWWD-AGRL: 0.88*; WSTR in %.: 0.93*; Nitrogen use in Kgs/ha.: 0.31*; Potash use in Kgs/ha.: 0.84*; Phosphate use in Kgs/ha.: 0.74*; NPK use in kgs/ha.: 0.71*; FW/CAP in M3.: −0.98*; GDP/capita/year in US$.: 0.96*
P2 (1990–2000)	None	TCS/day/person in Kcal.: 0.89*; SAP in %.: 0.95*; RH in %.: 0.53**; % AL.: 0.94*; %FWWD-AGRL.: 1.00*; Nitrogen use in Kgs/ha.: 0.94*; Potash use in Kgs/ha.: 0.92*; Phosphate use in Kgs/ha.: 0.97*; NPK use in kgs/ha.: 0.98*; FW/CAP in M3.: −0.99*; GDP/capita/year in US$.: 1.00*	SAP in %.: 0.83*; RH in %.: 0.59**; % AL.: 0.87*; FC in %.: 0.89*; %FWWD-AGRL.: 0.88*; Pesticide use in Kgs/ha.: −0.55**; Nitrogen use in Kgs/ha.: 0.82*; Potash use in Kgs/ha.: 0.88*; Phosphate use in Kgs/ha.: 0.78*; NPK use in kgs/ha.: 0.86*; FW/CAP in M3.: −0.87*; GDP/capita/year in US$.: 0.86*
P3 (2001–2010)	None	None	None
P4 (2011–2020)	% MP.: −0.74*; RH in %.: −0.56**; % AL.: 0.80*; FC in %.: 0.77*; Pesticide use in Kgs/ha.: 0.56**; Nitrogen use in Kgs/ha.: 0.56**; Potash use in Kgs/ha.: −0.50**; FW/CAP in M3.: – 0.77*; GDP/capita/year in US$.: 0.73*	% MP.: −0.99*; % AL.: 0.91*; AATP Anomaly in C: 0.77*; Pesticide use in Kgs/ha.: 0.57*; Phosphate use in Kgs/ha.: −0.41***; FW/CAP in M3.: −1.00*; GDP/capita/year in US$.: 0.99*	None

Note *, **, and *** indicate the level of significance at 1%, 5%, and 10% levels, respectively
Source Estimated by the authors

3.4 Nature

The number of indicators has shown a positive association with nature (forest cover). These include: food (total calorie supply), share of animal protein, temperature anomaly, agricultural land, agricultural water use, water stress, fertilizer use, pesticide use, and GDP per capita (Table 5). Over the past 30 years, temperature anomalies have complemented forest cover. Forest cover and food (calorie supply) are positively linked, and the share of animal protein is also positively linked to forest cover. This indicates that forests complement the food needs (forest products). It also supports the earlier observations that forests support the production of animal protein. The positive association with water stress, water use for agriculture, and chemical use (NPK, pesticides, etc.) is not straightforward and needs further investigation, which is detailed in the next section. The share of agricultural land and water stress is expected to move in the opposite direction to forest cover. The positive association between forest cover and GDP per capita is expected, especially given the policy focus on afforestation programs in Vietnam. On the other hand, forest cover has a negative association with the extent of malnutrition and water availability. Even the extent of malnutrition is inversely linked to forest cover, indicating biodiversity is not improving the nutrition status. Additionally, the signs of most indicators have been consistent over the years.

3.5 Food

Over the 30-year period, calorie supply/availability is positively associated with agricultural land, forest cover, share of animal protein, temperature anomaly, water use for agriculture, water stress, NPK use, and GDP per capita (Table 5). Nature indicators, such as forest cover and agricultural water use, along with intensification indicators, are positively and expectedly linked with total calorie availability. On the other hand, the negative association between the extent of malnutrition and total calorie availability is also consistent with theoretical expectations. In contrast, water availability per capita has turned out to be significant with a negative sign. This indicates that agriculture is prioritized for water allocation, as it is critical for food security. Most of the indicators have shown consistent association in the sub-period two as well. None of the indicators turned significant in periods three and four. The next section examines the more intricate and indirect linkages among the three dimensions in relation to Asia.

3.6 A Food-Insecure Developing Economy

Vietnam has overlaps with each of the other nations under analysis. It has a massive biological base, a rapidly growing and strong economy, and experiences periods of food security stress, including times of famine. Vietnam has a high correlation with total calorie availability and the share of animal protein (Table 7). It also has a very high correlation with the share of agricultural land and a high correlation with total calorie availability. Interestingly, the share of agricultural land has a high correlation with both total calorie availability and the share of animal protein, which is contrary to other nations.

Vietnam has adopted a very aggressive approach to agricultural intensification in order to alleviate acute food security challenges, which has been balanced by its existing large natural resource base, abundant in both forest cover and rainfall. However, Vietnam's economy has also been greatly bolstered by the rapid adoption and expansion of manufacturing, particularly in the textile industry, leading to widespread pollution and the rapid degradation of land and water resources.

Vietnam also has a poor correlation of annual precipitation to total calories (0.10), indicating that agriculture is largely supported by irrigation infrastructure, including groundwater. However, continued pollution and degradation of these natural resource bases continue to enhance the reliance on intensification. While earlier intensification measures were largely focused on expanding productivity to meet growing needs, this activity is now more focused on sustaining the high productivity level by expanding into marginal lands and water resources, as evidenced by the shift in correlations over the last decade (Table 7). Vietnam is a case study in how poor management of the C-N-F Nexus may erode any advantages of having a rich natural resource base, and mismanagement of these resources may downgrade the country to a stressed C-N-F status. The transition of correlations from the levels during 1990–2000 to the 2010–2020 levels carries significant meaning. Lessons from Japan could be key in guiding

Table 7 Correlation coefficients between food indicators and other indicators for Vietnam (1990–2020; 1990–2000 and 2011–2020)

	1990–2020		1990–2000		2011–2020	
Indicator	TCS/day/person in Kcal	SAP in %	TCS/day/person in Kcal	SAP in %	TCS/day/person in Kcal	SAP in %
SAP in %	0.98*	1	0.83*	1	0.30	1
% AL	0.98*	0.95*	0.87*	0.91*	0.80*	0.21
FC in %	0.99*	0.97	0.89*	0.95*	0.97*	0.23
%FWWD-AGRL	0.88*	0.85*	0.88*	0.94*	−6.43845E-16	1.85E-15
WSTR in %	0.93	0.94*	–	–	–	–
NPK use in Kgs/ha	0.71*	0.68*	0.86*	0.96*	−0.04	−0.44***
Pesticide use in Kgs/ha	–	–	−0.55**	−0.17	0.73*	0.14

Note *, **, and *** indicate the level of significance at 1%, 5%, and 10% levels, respectively
Source Estimated by the authors

economies like Vietnam back to pathways of better natural resource conservation, while continuing to boost and grow the economy.

Vietnam establishes itself very strongly; therefore, it is important to treat the C-N-F as a cohesive whole, and what are the risks of addressing any of the vertices of the nexus in isolation from the others? As a result, there is always a significant risk of exacerbating issues in other systems. In the absence of a comprehensive C-N-F strategy, it will tend to constrain the progress being made in the system of focus. That is, focusing solely on food systems will result in the degradation or aggravation of issues in both natural and climate systems, ultimately restricting progress in bolstering and expanding the food production system. An isolated focus on climate systems would entail the adoption of strategies that restrict the utilization of natural resources, which may jeopardize food productivity. Focusing on conservation of nature may build up constraints on land use, pest control, and climatic condition management, leading to disruptions in climate management and food production activities.

4 Conclusions

The case of Southeast Asia and Vietnam provides another interesting scenario for the C-N-F transition over the period that has evolved alongside economic transformation from agriculture to manufacturing and then to services. Unlike in the case of East and South Asia, there is no overbearing or dominant nation in the region, as all the member countries are small in size and demographics. Despite the homogeneity in size, resource endowments, climate risks, and the trajectory of economic transformation among the small nations of the region, they differ widely in their well-being, i.e., the extent of poverty, food security, GDP growth, etc.

Vietnam is a better-performing nation in the region, with consistent growth rates and a low incidence of poverty. This is achieved through a singular focus on food security through intensification and industrial growth, with little concern for biodiversity. While this has helped achieve food security and economic growth, it has come at the cost of natural systems, resulting in biodiversity loss and declining ecosystem services. Such a siloed approach is not sustainable in the long run. Unless a comprehensive and integrated approach to C-N-F is adopted, it is more likely to result in inefficient and undesirable outcomes. Vietnam offers valuable insights into why a linear approach, focusing on either of the three dimensions (C-N-F), is ineffective in addressing sustainable development.

Appendix

See Table 8.

Table 8 Correlation coefficients of the C-N-F nexus indicators for Vietnam

	1990–2020			1990–2000			2001–2010			2011–2020		
	TCS/day/person in Kcal	FC in %	AATP anomaly in C	TCS/day/person in Kcal	FC in %	AATP anomaly in C	TCS/day/person in Kcal	FC in %	AATP anomaly in C	TCS/day/person in Kcal	FC in %	AATP anomaly in C
TCS/day/person in Kcal	1.00	0.99*	0.57**	1.00*	0.89*	0.05	1E+00	1E+00	9E−03	1E+00	1E+00	9E−03
SAP in %	0.98*	0.97*	0.51**	0.83*	0.95*	0.07	1E+00	1E+00	7E−02	1E+00	0.24	−0.10
% MP	−0.95*	−0.93*	−0.59**	–	–	–	−1E+00	−1E+00	−1E−01	−1E+00	−0.99*	−0.74*
RH in %	−0.06	−0.04	−0.41**	0.59**	0.53**	−0.39	6E−02	1E−01	−5E−01	6E−02	−0.35	−0.56**
% AL	0.98*	0.99*	0.62**	0.87*	0.94*	0.10	9E−01	1E+00	1E−01	9E−01	0.91*	0.80*
FC in %	0.99*	1.00	0.58**	0.89*	1.00	0.13	1E+00	1E+00	6E−02	1E+00	1.00	0.77*
AARF anomaly in mm	0.10	0.10	−0.19	0.34	0.26	−0.33	6E−01	6E−01	−1E−01	6E−01	0.03	−0.34
AATP anomaly in C	0.57**	0.58**	1.00	0.05	0.13	1.00	9E−03	6E−02	1E + 00	9E−03	0.77	1.00
%FWWD−AGRL	0.88*	0.92*	0.43**	0.88*	1.00*	0.11	8E−01	9E−01	−1E−01	8E−01	0.00	0.00
WSTR in %	0.93*	0.90*	0.46**	0.00	0.00	0.00	8E−01	9E−01	−1E−01	8E−01	–	–
Pesticide use in Kgs/ha	0.07	0.08	0.17	−0.55**	−0.25	−0.23	4E−01	4E−01	4E−01	4E−01	0.57**	0.56**
Nitrogen use in Kgs/ha	0.31***	0.39***	0.40***	0.82*	0.94*	0.25	−1E−01	−2E−01	4E−01	−1E−01	0.39	0.56**
Potash use in Kgs/ha	0.84*	0.88*	0.40***	0.88*	0.92*	0.23	2E−01	2E−01	−4E−01	2E−01	−0.31	−0.50**

(continued)

Table 8 (continued)

	1990–2020			1990–2000			2001–2010			2011–2020		
	TCS/day/person in Kcal	FC in %	AATP anomaly in C	TCS/day/person in Kcal	FC in %	AATP anomaly in C	TCS/day/person in Kcal	FC in %	AATP anomaly in C	TCS/day/person in Kcal	FC in %	AATP anomaly in C
Phosphate use in Kgs/ha	0.74*	0.78*	0.27	0.78*	0.97*	0.06	6E−01	7E−01	−8E−02	6E−01	−0.41***	−0.35
NPK use in kgs/ha	0.71*	0.78*	0.42**	0.86*	0.98*	0.21	3E−01	3E−01	2E−01	3E−01	0.06	0.15
FW/CAP in M3	−0.98*	−1.00*	−0.58**	−0.87*	−0.99*	−0.11	−1E+00	−1E+00	−6E−02	−1E+00	−1.00*	−0.77*
GDP/capita/year in US$	0.96*	0.95*	0.65*	0.86*	1.00*	0.16	1E+00	1E+00	5E−02	1E+00	0.99*	0.73*

Note *, **, and *** indicate the level of significance at 1%, 5%, and 10% levels, respectively
Source Estimated by the authors

References

Ha Van T, Lichang L, Dang Thanh Quoc T (2024) Sustainable development in Southeast Asia: the nexus of tourism, finance, and environment. Heliyon 10(24):e40829. (https://doi.org/10.1016/j.heliyon.2024.e40829. PMID: 39720077; PMCID: PMC11667621

Huang B, Xu Y (2019) Environmental performance in asia: overview, drivers, and policy implications. In: ADBI working paper 990. Asian Development Bank Institute, Tokyo. https://www.adb.org/publications/environmental-performance-asia-overview-drivers-policy-implications

Mendelsohn R (2014) The impact of climate change on agriculture in Asia. J Integr Agric 13(4):660–665. https://doi.org/10.1016/S2095-3119(13)60701-7

Raihan A, Ridwan M, Sarker T, Atasoy FG, Zimon G, Bari AM, Rahman MS, Khan H, Mohajan B (2025) The influence of different environmental factors toward Vietnam's net-zero emissions goal. Innov Green Dev 4(3):100229. https://doi.org/10.1016/j.igd.2025.100229

United Nations (2024) Vietnam common country analysis: 2024 update. United Nations, Vietnam, Hanoi

Part III
Managing the Nexus

Chapter 8
Tracking the Nexus: An Integrated Monitoring Index

1 Background

Managing the C-N-F nexus in a comprehensive and integrated manner addresses three important Sustainable Development Goals (SDGs): 2, 13.1, 13.2, and 15.2 targets. Goal 2 is an overarching global target of achieving "Zero Hunger" (ensuring the availability and access to sufficient and nutritious food for all). Achieving this efficiently and sustainably requires meeting the requirements of 13.1, 13.2, and 15.2. These (13.1, 13.2 and 15.2) are: (i) "Strengthen resilience and adaptive capacity to climate-related hazards and natural disasters in all countries"; (ii) "Integrate climate change measures into national policies, strategies and planning globally" and (iii) "Promote the implementation of sustainable management of all types of forests, halt deforestation, restore degraded forests and substantially increase afforestation and reforestation globally".[1] Proper management of the C-N-F nexus addresses climate and nature dimensions in a manner that complements each other and ensures the sustainability of food systems. The C-N-F nexus is a dynamic concept that changes spacio-temporally, as each dimension is represented by a set of core indicators connected to a number of other variables, both independent and dependent. As in the case of SDG goals, monitoring the C-N-F nexus helps track its progress or deterioration. Monitoring helps identify gaps and allocate limited resources effectively. Tracking the C-N-F nexus is more valuable in the context that the combined effect of the interactions between the three facets has a greater value than the individual contributions in isolation (SDG goals). This chapter develops an integrated C-N-F nexus monitoring index and constructs the indices for Asia, its four sub-regions, and the four selected countries from each sub-region.

Monitoring tools, such as indices, are required to understand the status and have a comparative picture over time and across geographies. Monitoring is also useful within the country to assess regional differences in performance and progress. While

[1] https://sdgs.un.org/goals/goal16.

V. R. Reddy et al., *Strengthening the Climate–Nature–Food Nexus*,
https://doi.org/10.1007/978-3-032-29423-4_8

targets are set at the global level, individual countries are responsible for achieving them. Monitoring helps in assessing their strengths and weaknesses in achieving the targets. Continuous monitoring helps determine the pace at which targets are being reached and identify measures to enhance it. Monitoring changes in indicators helps track progress and assess whether it is in the desired direction. Any deviations reflect the effectiveness or appropriateness of the interventions. Based on the requirements, international support can be provided in designing appropriate policy measures and interventions for the sustainable management of the C-N-F nexus. These measures could take the form of infrastructure development, capacity building, incentive/disincentive structures, and so on, that can directly influence the process. Indirectly, awareness-building efforts, incentives, subsidies, and other measures could help achieve the targets.

2 Approach

Based on the C-N-F dimensions and their indicators discussed in the previous chapters, four indices of the nexus are constructed, namely climate, nature, food, and the integrated C-N-F index (C-N-F I). These indices are constructed at the sub-region/region and individual country levels. The status at each level could be assessed component-wise and by an aggregated index. And, countries or regions can be ranked based on the aggregate performance. For the present analysis a few core indicators representing each dimension have been considered i.e., climate: temperature anomalies (TA), precipitation anomalies (PA) and relative humidity (RH); nature: % of forest cover (FC), % of agriculture land (AL) and water stress (WS); food: total calorie supply (TC), percentage of malnourished population (MP) and share of animal protein (AP). These are referred to as the Climate Component Index (CCI), the Nature Component Index (NCI), the Food Component Index (FCI), and the integrated Climate–Nature–Food Index (CNFI).

2.1 Normalization of Indicators

Often, the data on these indicators is available from different sources and formats. In the case of comparisons across time periods or geographies, the data needs to be normalized, as each quantified indicator may have different maximum and minimum values and different mean values. The distance function is most commonly used for normalizing the data in the construction of integrated indices. The indicator value of variable x, country/sub-region/region 'i' and time period 't' is computed as

$$d\ (x_{it}) = (\text{Actual}\, x_{it} - \text{Min}\, x_{it})/(\text{Max}\, x_{it} - \text{Min}\, x_{it})$$

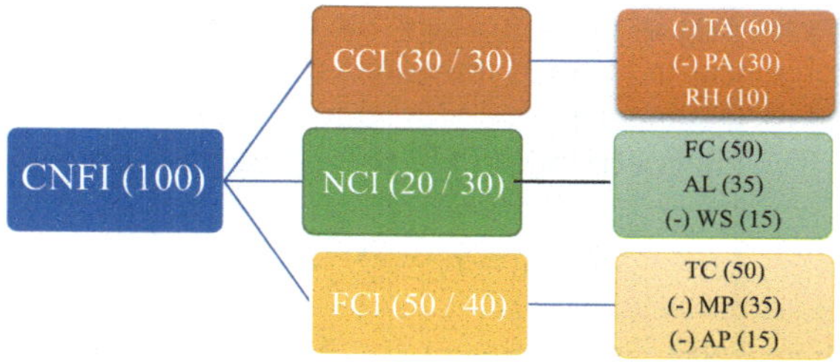

Fig. 1 Conceptual model for integrated C-N-F Index (CNF I). *Note* Figures in brackets are theoretically proposed signs and weights

where Min x_{it} is the minimum value of $x_{it,}$ and Max x_{it} is the maximum value of x_{it} observed over the years. Following the theoretical consideration and the correlation analysis in the previous chapters, signs are assigned to each indicator depending on their inter-linkages (Fig. 1). As a result, the index ranges between '−1 and + 1' indicating that below zero or negative values indicate negative impacts and + 1' indicate positive impacts. Although some of the indicators and components may have negative or positive signs, the integrated C-N-F index, depending on its sign, indicates the status of the nexus.

2.2 *Weights*

Once the data is normalized, weights are assigned to different indicators and dimensions depending on their importance. Several approaches are employed to determine the weights of different indicators (OECD 2008). In the present case, weights are determined based on the theoretical understanding and the analysis in the previous chapters. The conceptual model is presented in Fig. 1. Weights for the indicators representing each dimension are assigned (Fig. 1). Two different sets of weights are assigned at the component level to provide scenarios. A normal or weighted index can be constructed depending on the context; however, a weighted index is more robust. Due to the assignment of signs to each indicator, only weighted indices are presented here.

Index for component 'k'; region/country 'j' and time 't' can be denoted as,

$$I_{kjt} = \sum\nolimits_i W_{it}\, d_{ijt},\ \forall\, \text{all}\, d_i \in \text{Componenent}\, k \sum W_{it}$$

where Wit is the weights for the indicators concerned, and d_{ijt} is the ith indicator for jth region/country in time 't' for component 'k'. In this case, each indicator gets its assigned weight in the component index.

For the component-wise aggregate Index for component 'k' is,

$$I_{k\text{t}} = \sum\nolimits_j W_j I_{kjt} = \text{CCI}_{jt}/\text{NCI}_{jt}/\text{FCI}_{jt}$$

where '*Wj*' is the weightage for the component concerned. CCI_{jt} is the climate component index of '*j*'th region/country at time '*t*'; NCI_{jt} and FCI_{jt} are the nature and food components of '*j*'th region/country in time '*t*'.

The integrated index of C-N-F nexus for region/country '*j*' is

$$\mathrm{CNF}I_{jt} = I_{jt} = \sum I_{kjt} W_{\mathrm{kjt}} + \sum I_{ljt} W_{\mathrm{ljt}} + \sum I_{ljt}\, W_{ljt}$$

where $\mathrm{CNF}I_{jt}$ is the integrated C-N-F nexus index for region/country '*j*' in time '*t*'.

I_{kjt} is the index of component '*k*' for components: CCI, NCI, and FCI.

3 Construction of C-N-F I: Spacio-temporal Indices

An integrated index of climate, nature, and food components, along with their corresponding indicators, is constructed using the data presented in the previous chapters. To make it simple, only nine indicators are included in constructing the CNF_{Iit}, i.e., three indicators for each component. Different weights are assigned to each indicator (Fig. 1). The CCI component is given a 30% weight, the NCI component is given a 20% weight, and the FCI component is given a 50% weight. Alternative weights of 30 (CCI), 30 (NCI), and 40 (FCI) are also assigned for the purpose of scenario analysis. In fact, weights are fixed for each indicator first and then summed up by component. Indices are constructed for the three time periods, i.e., using the averages of the 31-year data (1990–2020), and for the first decade (1990–2000) and the latest decade (2011–2020). While the more extended period indices, along with indicator-wise details, are presented in the appendix, here we discuss the first and latest decades to examine any temporal shifts in the status of the nexus indices. It may be noted that the construction of indices is constrained by the same limitations (as mentioned in Chap. 2), such as missing or discontinuous data for some indicators, which create anomalies in the normalization process. Indices are constructed for Asia and its four sub-regions and for the four selected countries. The integrated C-N-F indices are presented here; the detailed component- and indicator-wise indices are provided in the appendix.

3.1 Discussion

Thus, the scale has been defined between −1 and 1. "Zero" indicates the ideal balance between nexus elements, while −1 means extreme degradation of all three facets, and +1 indicates an aggressive conservatory approach, which is unsustainable in the long run. This is mainly because of the trade-off between the indicators. For instance, forest cover, which has a positive influence, may impact the extent of agricultural land (positive influence) adversely beyond a point. Such trade-offs, in turn, would impact food production. Similarly, relative humidity that is expected to have a positive influence is likely to have a negative impact beyond a threshold level. Thus, ecological systems tend to be unsustainable as the indices approach to '+1'. The index value for the levels considered, however, is close to 0, meaning there is no extreme case of conservation or degradation. However, the values are still significant in terms of measuring how well the nexus is managed.

The index for the last decade is more valuable as an indicator of the current state of the nexus and the areas that may need to be focused on to achieve a better balance. The last decade index value can be compared with previous time frame indices in order to measure the progress made. For the purpose of this discussion, two decadal indices are used for comparative analysis: 1990–2000 and 2011–2020. The scales for the indices are presented for the 30-year period for the sub-regions as well as for the selected countries.

3.2 *Asia Indices*

Three scenarios are constructed for each time period for the Asia aggregate as well as for each sub-region. Scenario I (S-I) represents weighted indices at the indicator level but not at the component level (partial weightage). Scenario II (S-II) represents weighted indices at both the indicator and component levels. Scenario III (S-III) uses alternative weights at the component level. Given the variations in the C-N-F dimensions across the regions, S-II and S-III are more relevant. At the Asia level, the integrated C-N-F indices have shown negative values during the first period (1990–2000). All the indices suggest a borderline status, either negative or positive. However, they have turned positive over the recent decade, indicating some improvement in the status of the C-N-F nexus in recent years (Table 1). The improvement in the nexus status is quite reasonable, especially in S-I and S-II.

The improvement over the period is mainly due to the food component, which has reported a substantial improvement. Between the periods, the food component has shown a clear improvement (from negative to positive), while the nature component

Table 1 Integrated CNF index for Asia

Component/period	1990–2000				2011–2000			
	CCI	NCI	FCI	C-N-F I	CCI	NCI	FCI	C-N-F I
Indicator weighted: SI	−0.39	0.40	−0.23	−0.22	−0.31	0.39	0.05	0.14
Component weighted (30:20:50): S-II	−0.12	0.08	−0.12	−0.16	−0.09	0.08	0.01	0.00
Component weights (30:30:40): S-III	−0.12	0.12	−0.09	−0.09	−0.09	0.12	0.02	0.05

Note CCI = Climate component Index; NCI = Nature component index; FCI = Food component index; C-N-F I = Climate, nature, food integrated index

remains stable, albeit on the positive side. The climate component has shown some improvement between the periods, although it remains negative. It is observed that food availability in most Asian countries has improved over the period. Besides, forestry (nature) has received strong policy support in most of the Asian countries, especially in big countries like India. In terms of individual indicators, forest cover has declined between the two periods (see Appendix Table 4). On the contrary, water resources or water stress declined, which is mainly due to the development of water resources in most Asian countries. The share of agricultural land has increased marginally due to the trade-off between forest cover and agricultural land. The positive linkages between forest cover and agricultural land, as well as the negative association with water stress, are well reflected in the status quo of the nature component between the periods. In the case of food, all the indicators have reported improvement between the periods (Appendix Table 4).

3.3 Sub-regional Indices

The sub-regional indices bring out differences between the sub-regions in terms of C-N-F status. East and Southeast Asia are clearly positive in both periods, though Southeast Asia has turned negative during the recent decade in S-II (Table 2). On the other hand, central and South Asia are in negative territory, especially during the recent decade. Of the sub-regions, East Asia is better placed, followed by Southeast Asia, South Asia, and Central Asia. Of the sub-regions, Central Asia indices turned from positive to negative between the periods, indicating a deterioration in nexus status. This is mainly due to the decline in the status of the nature and food dimensions, although the status of the climate dimension has improved. All the other regions revealed an improvement in the nexus status over the period. The nature component is observed to be positive in all cases, and the climate component is negative. It may be noted that the nature component has either deteriorated or remained stable in most cases. In fact, Central Asia has shown a decline in the value of the nature component indices. On the other hand, the climate component has shown some clear improvement in the majority of cases. All the sub-regions, except Central Asia, have improved the status of the food component from negative to positive between the two periods. Although Central Asia has positive food indices (S-I and S-II) during both periods, they have recorded a substantial decline in the recent decade (Table 2).

These changes suggest that Asia and its sub-regions, except Central Asia, appear to be managing the nexus at present, which may be due to the current state of the nature component. Central Asia has a declining forest cover and increasing water stress (Appendix Table 5). While most regions are trying to maintain their forest resources, degradation of water resources is widespread across the region. At the same time, the climate component has not yet reached a tipping point in the sub-regions. In fact, temperature anomalies are within the range that supports food systems in Asia (Mendelsohn 2014). The current approaches are insufficient to build the resilience of systems in terms of enhancing biodiversity and ecosystem services. This is evident

Table 2 Integrated CNF index for the sub-regions of Asia

Component/period	1990–2000				2011–20			
	CCI	NCI	FCI	C-N-F index	CCI	NCI	FCI	C-N-F index
Central Asia								
Indicator weighted—S-I	−0.41	0.24	−0.29	−0.46	−0.29	0.20	0.03	−0.06
Component weighted (30:20:50)—S-II	−0.12	0.05	0.15	0.08	−0.09	0.04	0.02	−0.06
Component weights (30:30:40)—S-III	−0.12	0.07	0.12	0.07	−0.09	0.06	0.01	−0.02
East Asia								
Indicator weighted—S-I	−0.42	0.42	0.10	0.10	−0.30	0.38	0.27	0.35
Component Weighted (30:20:50)—S-II	−0.13	0.08	0.05	0.00	−0.09	0.08	0.14	0.13
Component weights (30:30:40)—S-III	−0.13	0.13	0.04	0.04	−0.09	0.11	0.11	0.13
South Asia								
Indicator weighted—S-I	−0.39	0.42	−0.26	−0.23	−0.31	0.47	−0.14	0.02
Component weighted (30:20:50)—S-II	−0.12	0.08	−0.13	−0.17	−0.09	0.09	−0.07	−0.07
Component weights (30:30:40)—S-III	−0.12	0.13	−0.10	−0.09	−0.09	0.14	−0.06	−0.01
South East Asia								
Indicator weighted-S-I	−0.06	0.49	−0.28	0.15	−0.11	0.50	0.07	0.46
Component weighted (30:20:50)—S-II	−0.02	0.10	−0.14	−0.06	−0.03	0.01	0.04	0.02
Component weights (30:30:40)—S-III	−0.02	0.15	−0.11	0.02	−0.03	0.15	0.03	0.15

Note CCI = Climate component Index; NCI = Nature component index; FCI = Food component index; C-N-F I = Climate, nature, food integrated index

from the meagre or no improvement in the nature component in most cases. While public policies are more proactive in supporting food security (food component), they are more reactive in the case of climate management. The nature component, which enhances the system's resilience to climate risks, is not receiving due attention. Although these regions are currently in balance, they are not doing enough to prepare for future shocks. The neglect of the nature component may exacerbate vulnerabilities. One can observe this process at work in the cases of Central and South Asia. The country-level indices in the following section may provide further insights in this regard.

3.4 Country Indices

Country indices reveal an opposite picture when compared to the sub-regional indices. This is mainly due to the reason that the selected countries do not well represent the sub-region, and some of the sub-region aggregates are dominated by large countries, such as Pakistan (Central Asia), India (South Asia), and China (East Asia). All the selected countries, except Uzbekistan, have been in negative territory during the recent decade. In the recent decade, all the countries except Uzbekistan have turned to negative (Table 3). The case of Uzbekistan is intriguing, as it turned from negative to positive in all three scenarios. This has primarily resulted from the food component. In recent years, Uzbekistan has reported the highest total calorie supply (Chap. 4) compared to the other three countries and Central Asia as well. Uzbekistan has maintained its status in the case of the natural component and has experienced marginal changes in the case of the climate component. This clearly indicates that the improvement in the C-N-F status in Uzbekistan is primarily attributed to policy changes that support food security. All the indicators of food performance were well above average over the period (Appendix Table 9).

Interestingly, Japan's status has shifted from positive to negative, indicating an accelerating degradation of its natural environment, food systems, and climate, especially over the recent decade in all three scenarios. Japan also exhibited a contrasting status compared to East Asia. While at the sub-regional level, East Asia showed a positive status in all scenarios in both periods, Japan revealed a negative status in most cases (Table 3). This is mainly due to the drastic decline in the nature component over the period. Among the components, the nature component has shown a substantial decline in its indices (status) between the periods, while the climate component has shown an improvement, and the food component has shown a marginal improvement. At the indicator level, forest cover has declined substantially between the periods (Appendix Table 10). This is mainly due to the increased demand for land for other uses. It appears that Japan places more emphasis on climate management while neglecting the nature component, which may prove detrimental in the coming years. Climate management needs to be complemented by improved biodiversity and ecosystems for the sustained resilience of the system and people in addressing

Table 3 Integrated CNF index for the selected countries in Asia

Components/ period	1990–2000				2011–2020			
	CCI	NCI	FCI	C-N-F I	CCI	NCI	FCI	C-N-F I
Uzbekistan								
Indicator weighted-S-I	−0.09	0.21	−0.05	0.07	−0.40	0.22	0.36	0.18
Component weighted (30:20:50)—S-II	−0.03	0.04	−0.03	-0.02	−0.12	0.04	0.18	0.10
Component weights (30:30:40)—S-III	−0.03	0.06	−0.02	0.01	−0.12	0.06	0.14	0.08
Japan								
Indicator weighted-S-I	−0.41	0.49	0.23	0.31	−0.15	0.13	0	−0.02
Component weighted (30:20:50)—S-II	−0.12	0.10	0.12	0.10	−0.05	0.03	0	−0.02
Component weights (30:30:40)—S-III	−0.12	0.15	0.09	0.12	−0.05	0.04	0	−0.01
Nepal								
Indicator weighted-S-I	−0.40	0.38	−0.29	0.27	−0.33	0.40	−0.08	−0.01
Component weighted (30:20:50)—S-II	−0.12	0.08	−0.15	−0.19	−0.10	0.08	−0.04	−0.003
Component weights (30:30:40)—S-III	−0.12	0.11	−0.12	−0.13	−0.10	0.12	−0.03	−0.003
Vietnam								
Indicator weighted-S-I	−0.39	0.28	−0.23	−0.34	−0.40	0.49	−0.09	0.0
Component weighted (30:20:50)—S-II	−0.12	0.06	−0.12	−0.18	−0.12	0.10	−0.05	−0.07
Component weights (30:30:40)—S-III	−0.12	0.08	−0.09	−0.13	−0.12	0.15	−0.04	−0.01

Note CCI = Climate component Index; NCI = Nature component index; FCI = Food component index; C-N-F I = Climate, nature, food integrated index

climate risks. Despite being the richest country in Asia, it is unable to balance the nexus and move towards sustainability.

Interestingly, Nepal's indices seem to be closer to the sub-regional status, despite the presence and influence of a big country like India in the sub-region. The status of the C-N-F nexus is negative in both periods, with marginal changes in most cases. Only in the case of the climate component is there some improvement. Even the nature indicators have remained stable. There is also good improvement in the case of food, while the climate component remained stable. This corroborates the understanding that Nepal has prioritized nature and climate over food security, leading to poorer caloric outcomes in the country. It illustrates that, even though Nepal appears to be doing well in terms of conservation, it still does not indicate a well-balanced nexus. It is insufficient to manage only two facets of the nexus while keeping the third as a secondary priority.

Vietnam is another case where country indices show the opposite of the sub-regional indices. Indices are negative in both periods, although there has been an improvement in the recent decade. The nature component has shown good improvement in all three scenarios. This appears to have a positive influence on both the climate and food dimensions, as both components showed improvement over the period. Among the natural indicators, forest cover and agricultural land have revealed substantial improvement, while water stress has remained the same (Appendix Table 12). This also explains the increased food availability. At the same time, the share of animal protein has also improved.

4 Conclusions

Integrated indices help determine the status of the C-N-F nexus across regions and countries between two time periods. Indices clearly indicate that most sub-regions and countries are on the margins (close to zero) in terms of status, although they are in the negative in the majority of cases. There are variations across the sub-regions and countries. It is observed that most sub-regions and countries have focused on food security or the food component to the neglect of the nature component. The focus on food security is more pronounced in the case of the resource-poor Uzbekistan. The neglect of the nature component is evident even in the case of Japan, which is the richest country in Asia. While most of the sub-regions and countries are maintaining or protecting their forest resources, all of them are experiencing water resource degradation. The continuation of the water resource degradation process or increased water stress directly impacts food production. In fact, climate change impacts food systems through water systems. These impacts are evident at the micro level, although they have yet to be reflected at the aggregate level. Thus, the indices indicate that the C-N-F is not moving in a balanced or desired direction, which may impact the resilience of the natural system. While focusing on food security helps address the immediate priorities of hunger and poverty, the imbalance between nature and food could have an adverse impact on food systems in the medium to long term. Neglect of the natural systems would aggravate the pace and intensity of climate risks. Such adverse impacts are yet to be conspicuous at the aggregate level to prompt policy action and monetary support. Most countries seem to be focusing on quick fixes for addressing climate risks rather than devising comprehensive approaches to build and strengthen resilience through investments and financing of natural capital or biodiversity initiatives.

Appendices

See Tables 4, 5, 6, 7, 8, 9, 10, 11 and 12.

Table 4 Component wise indices for Asia

	1990–2000			2011–20			1990–2020		
	Actual value	Weight	Weighted value	Actual value	Weight	Weighted value	Actual value	Weight	Weighted value
Climate									
TA	−0.49	0.6	−0.29	−0.47	0.60	−0.28	−0.49	0.6	−0.29
PA	−0.61	0.3	−0.18	−0.39	0.30	−0.12	−0.62	0.3	−0.19
RH	0.91	0.1	0.09	0.90	0.10	0.09	0.90	0.1	0.09
CCI	−0.19	1	−0.39		1	−0.31		1	−0.39
Nature									
FC	0.60	0.50	0.30	0.54	0.5	0.27	0.57	0.5	0.29
AL	0.51	0.35	0.18	0.53	0.35	0.19	0.55	0.35	0.19
WS	−0.54	0.15	−0.08	−0.42	0.15	−0.06	−0.5	0.15	−0.08
NCI	0.57	1	0.40		1	0.39		1	0.40
Food									
TC	0.24	0.5	0.12	0.54	0.5	0.27	0.45	0.5	0.23
MN	−0.87	0.35	−0.30	−0.44	0.35	−0.15	−0.58	0.35	−0.20
AP	−0.31	0.15	−0.05	−0.45	0.15	−0.07	−0.39	0.15	−0.06
FCI	−0.94	1	−0.23		1	0.05		1	−0.03

Note CCI = Climate component Index; NCI = Nature component index; FCI = Food component index; C-N-F I = Climate, nature, food integrated index. TA = Temperature anomaly; PA = Precipitation anomaly; RH = Relative humidity; FC = Forest cover; AL = Agriculture land; WS = Water Stress; TC = Total calorie supply; MN = extent of malnutrition; and AP = Share of animal protein

Table 5 Component wise indices for Central Asia

	1990–2000			2011–20			1990–2020		
	Actual value	Weight	Weighted value	Actual value	Weight	Weighted value	Actual value	Weight	Weighted value
Climate									
TA	−0.49	0.6	−0.294	−0.47	0.6	−0.28	−0.49	0.6	−0.29
PA	−0.64	0.3	−0.192	−0.29	0.3	−0.09	−0.62	0.3	−0.19
RH	0.78	0.1	0.078	0.75	0.1	0.08	0.77	0.1	0.08
CCI		1	−0.408		1	−0.29		1	-0.40
Nature									
FC	0.09	0.5	0.05	0.002	0.5	0.001	0.1	0.5	0.05
AL	0.95	0.35	0.33	0.97	0.35	0.34	0.97	0.35	0.34
WS	−0.93	0.15	−0.14	−0.97	0.15	−0.15	−0.95	0.15	−0.14
NCI		1	0.24		1	0.20		1	0.25
Food									
TC	0.18	0.5	0.09	0.54	0.5	0.27	0.42	0.5	0.21
MN	−0.94	0.35	−0.33	−0.48	0.35	−0.17	−0.61	0.35	−0.21
AP	−0.32	0.15	−0.05	−0.48	0.15	−0.07	−0.41	0.15	−0.06
FCI		1	−0.29		1	0.03		1	−0.07

Note CCI = Climate component Index; NCI = Nature component index; FCI = Food component index; C-N-F I = Climate, nature, food integrated index. TA = Temperature anomaly; PA = Precipitation anomaly; RH = Relative humidity; FC = Forest cover; AL = Agriculture land; WS = Water Stress; TC = Total calorie supply; MN = extent of malnutrition; and AP = Share of animal protein

Table 6 Component wise indices for East Asia

	1990–2000			2011–20			1990–2020		
	Actual value	Weight	Weighted value	Actual value	Weight	Weighted value	Actual value	Weight	Weighted value
Climate									
TA	−0.49	0.6	−0.29	−0.47	0.6	−0.28	−0.49	0.6	−0.29
PA	−0.69	0.3	−0.21	−0.34	0.3	−0.10	−0.62	0.3	−0.19
RH	0.86	0.1	0.09	0.82	0.1	0.08	0.84	0.1	0.08
CCI		1	−0.42		1	−0.30		1	−0.40
Nature									
FC	0.71	0.5	0.36	0.75	0.5	0.38	0.72	0.5	0.36
AL	0.41	0.35	0.14	0.23	0.35	0.08	0.37	0.35	0.13
WS	−0.5	0.15	−0.08	−0.48	0.15	−0.07	−0.5	0.15	−0.08
NCI		1	0.42		1	0.39		1	0.41
Food									
TC	0.64	0.5	0.32	0.88	0.5	0.44	0.77	0.5	0.39
MN	−0.41	0.35	−0.14	−0.12	0.35	−0.04	−0.3	0.35	−0.11
AP	−0.5	0.15	−0.08	−0.88	0.15	−0.13	−0.77	0.15	−0.12
FCI		1	0.10		1	0.27		1	0.17

Note CCI = Climate component Index; NCI = Nature component index; FCI = Food component index; C-N-F I = Climate, nature, food integrated index. TA = Temperature anomaly; PA = Precipitation anomaly; RH = Relative humidity; FC = Forest cover; AL = Agriculture land; WS = Water Stress; TC = Total calorie supply; MN = extent of malnutrition; and AP = Share of animal protein

Table 7 Component wise indices for South Asia

	1990–2000			2011–20			1990–2020		
	Actual value	Weight	Weighted value	Actual value	Weight	Weighted value	Actual value	Weight	Weighted value
Climate									
TA	−0.49	0.6	−0.29	−0.47	0.6	−0.28	−0.49	0.6	−0.29
PA	−0.63	0.3	−0.19	−0.41	0.3	−0.12	−0.63	0.3	−0.19
RH	0.97	0.1	0.10	0.98	0.1	0.10	0.98	0.1	0.10
CCI		1	−0.39		1	−0.31		1	−0.39
Nature									
FC	0.62	0.5	0.31	0.65	0.5	0.33	0.63	0.5	0.32
AL	0.61	0.35	0.21	0.58	0.35	0.20	0.62	0.35	0.22
WS	−0.68	0.15	−0.10	−0.36	0.15	−0.05	−0.56	0.15	−0.08
NCI		1	0.42		1	0.47		1	0.45
Food									
TC	0.07	0.5	0.04	0.24	0.5	0.12	0.26	0.5	0.13
MN	−0.85	0.35	−0.30	−0.71	0.35	−0.25	−0.65	0.35	−0.23
AP	0.01	0.15	0.002	−0.05	0.15	−0.01	−0.06	0.15	−0.01
FCI		1	−0.26		1	−0.14		1	−0.11

Note CCI = Climate component Index; NCI = Nature component index; FCI = Food component index; C-N-F I = Climate, nature, food integrated index. TA = Temperature anomaly; PA = Precipitation anomaly; RH = Relative humidity; FC = Forest cover; AL = Agriculture land; WS = Water Stress; TC = Total calorie supply; MN = extent of malnutrition; and AP = Share of animal protein

Table 8 Component wise indices for South East Asia

	1990–2000			2011–20			1990–2020		
	Actual value	Weight	Weighted value	Actual value	Weight	Weighted value	Actual value	Weight	Weighted value
Climate									
TA	−0.003	0.60	−0.002	−0.01	0.6	−0.01	−0.01	0.6	−0.004
PA	−0.42	0.30	−0.13	−0.55	0.3	−0.17	−0.49	0.3	−0.15
RH	0.73	0.10	0.07	0.58	0.1	0.06	0.73	0.1	0.07
CCI		1	−0.06		1	−0.11		1	−0.08
Nature									
FC	0.98	0.50	0.49	0.98	0.5	0.49	0.95	0.5	0.48
AL	0.01	0.35	0.004	0.04	0.35	0.01	0.08	0.35	0.03
WS	0.00	0.15	0.00	−0.01	0.15	−0.001	−0.03	0.15	−0.005
NCI		1	0.49		1	0.50		1	0.50
Food									
TC	0.22	0.50	0.11	0.55	0.5	0.28	0.45	0.5	0.23
MN	−1.00	0.35	−0.35	−0.42	0.35	−0.15	−0.66	0.35	−0.23
AP	−0.25	0.15	−0.04	−0.4	0.15	−0.06	−0.35	0.15	−0.05
FCI		1	−0.28		1	0.07		1	−0.06

Note CCI = Climate component Index; NCI = Nature component index; FCI = Food component index; C-N-F I = Climate, nature, food integrated index. TA = Temperature anomaly; PA = Precipitation anomaly; RH = Relative humidity; FC = Forest cover; AL = Agriculture land; WS = Water Stress; TC = Total calorie supply; MN = extent of malnutrition; and AP = Share of animal protein

Table 9 Component wise indices for Uzbekistan (Central Asia)

	1990–2000			2011–20			1990–2020		
	Actual value	Weight	Weighted value	Actual value	Weight	Weighted value	Actual value	Weight	Weighted value
Climate									
TA	−0.43	0.6	−0.26	−0.56	0.6	−0.34	−0.62	0.6	−0.37
PA	0.52	0.3	0.16	−0.3	0.3	−0.09	−0.5	0.3	−0.15
RH	0.17	0.1	0.02	0.05	0.1	0.01	0.12	0.1	0.01
CCI		1	−0.09		1	−0.4		1	−0.51
Nature									
FC	0.004	0.5	0.002	0.01	0.5	0.01	0.02	0.5	0.01
AL	0.99	0.35	0.35	0.99	0.35	0.35	0.92	0.35	0.32
WS	−0.9	0.15	−0.14	−0.89	0.15	−0.13	−0.88	0.15	−0.13
NCI		1	0.21		1	0.22		1	0.20
Food									
TC	0.53	0.5	0.27	0.9	0.5	0.45	0.62	0.5	0.31
MN	−0.72	0.35	−0.25	0	0.35	0	−0.40	0.35	−0.21
AP	−0.39	0.15	−0.06	−0.57	0.15	−0.09	−0.46	0.15	−0.03
FCI		1	−0.05		1	0.36		1	−0.02

Note CCI = Climate component Index; NCI = Nature component index; FCI = Food component index; C-N-F I = Climate, nature, food integrated index. TA = Temperature anomaly; PA = Precipitation anomaly; RH = Relative humidity; FC = Forest cover; AL = Agriculture land; WS = Water Stress; TC = Total calorie supply; MN = extent of malnutrition; and AP = Share of animal protein

Table 10 Component wise indices for Japan (East Asia)

	1990–2000			2011–20			1990–2020		
	Actual value	Weight	Weighted value	Weight	Weighted value	Actual value	Actual value	Weight	Weighted value
Climate									
TA	−0.57	0.6	−0.34	−0.19	0.6	−0.11	−0.63	0.6	−0.38
PA	−0.49	0.3	−0.15	−0.19	0.3	−0.06	−0.51	0.3	−0.15
RH	0.76	0.1	0.08	0.19	0.1	0.02	0.74	0.1	0.07
CCI		1	−0.41		1	−0.15		1	−0.46
Nature									
FC	0.99	0.5	0.50	0.19	0.5	0.10	1	0.5	0.5
AL	0.05	0.35	0.02	0.19	0.35	0.07	0.03	0.35	0.01
WS	−0.21	0.15	−0.03	−0.19	0.15	−0.03	−0.19	0.15	−0.03
NCI		1	0.49		1	0.13		1	0.48
Food									
TC	0.74	0.5	0.37	0.19	0.5	0.10	0.66	0.5	0.33
MN	0	0.35	0	−0.19	0.35	−0.07	−0.005	0.35	−0.21
AP	−0.96	0.15	−0.14	−0.19	0.15	−0.03	−0.98	0.15	−0.03
FCI		1	0.23		1	0		1	−0.02

Note CCI = Climate component Index; NCI = Nature component index; FCI = Food component index; C-N-F I = Climate, nature, food integrated index. TA = Temperature anomaly; PA = Precipitation anomaly; RH = Relative humidity; FC = Forest cover; AL = Agriculture land; WS = Water Stress; TC = Total calorie supply; MN = extent of malnutrition; and AP = Share of animal protein

Table 11 Component wise indices for Nepal (South Asia)

	1990–2000			2011–20			1990–2020		
	Actual value	Weight	Weighted value	Actual value	Weight	Weighted value	Actual value	Weight	Weighted value
Climate									
TA	−0.56	0.6	−0.34	−0.45	0.6	−0.27	-0.62	0.6	−0.37
PA	−0.45	0.3	−0.14	−0.45	0.3	−0.14	-0.51	0.3	−0.15
RH	0.68	0.1	0.07	0.72	0.1	0.07	0.69	0.1	0.07
CCI		1	−0.40		1	−0.33		1	−0.46
Nature									
FC	0.54	0.5	0.27	0.56	0.5	0.28	0.56	0.5	0.28
AL	0.32	0.35	0.11	0.33	0.35	0.12	0.3	0.35	0.11
WS	0	0.15	0.00	0	0.15	0	−0.0002	0.15	−0.00003
NCI		1	0.38		1	0.40		1	0.39
Food									
TC	0.12	0.5	0.06	0.26	0.5	0.13	0.34	0.5	0.17
MN	−1.0	0.35	−0.35	−0.6	0.35	−0.21	−0.63	0.35	−0.21
AP	−0.01	0.15	0.00	−0.03	0.15	−0.01	−0.04	0.15	−0.03
FCI		1	−0.29						

Note CCI = Climate component Index; NCI = Nature component index; FCI = Food component index; C-N-F I = Climate, nature, food integrated index. TA = Temperature anomaly; PA = Precipitation anomaly; RH = Relative humidity; FC = Forest cover; AL = Agriculture land; WS = Water Stress; TC = Total calorie supply; MN = extent of malnutrition; and AP = Share of animal protein

Table 12 Component wise indices for Vietnam (South East Asia)

	1990–2000			2011–20			1990–2020		
	Actual value	Weight	Weighted value	Actual value	Weight	Weighted value	Actual value	Weight	Weighted value
Climate									
TA	−0.54	0.6	−0.32	−0.57	0.6	−0.34	−0.62	0.6	−0.37
PA	−0.52	0.3	−0.16	−0.36	0.3	−0.11	−0.51	0.3	−0.15
RH	0.95	0.1	0.10	0.96	0.1	0.10	0.94	0.1	0.09
CCI		1	−0.39		1	−0.40		1	−0.43
Nature									
FC	0.44	0.5	0.22	0.62	0.5	0.31	0.54	0.5	0.27
AL	0.19	0.35	0.07	0.53	0.35	0.19	0.34	0.35	0.12
WS	−0.05	0.15	−0.01	−0.06	0.15	−0.01	−0.05	0.15	−0.01
NCI		1	0.28		1	0.49		1	0.38
Food									
TC	0.12	0.5	0.06	0.45	0.5	0.23	0.43	0.5	0.22
MN	−0.79	0.35	−0.28	−0.71	0.35	−0.25	−0.54	0.35	−0.21
AP	−0.11	0.15	−0.02	−0.47	0.15	−0.07	−0.31	0.15	−0.03
FCI		1	−0.23		1	−0.09		1	−0.02

Note CCI = Climate component Index; NCI = Nature component index; FCI = Food component index; C-N-F I = Climate, nature, food integrated index. TA = Temperature anomaly; PA = Precipitation anomaly; RH = Relative humidity; FC = Forest cover; AL = Agriculture land; WS = Water Stress; TC = Total calorie supply; MN = extent of malnutrition; and AP = Share of animal protein

References

Mendelsohn R (2014) The impact of climate change on agriculture in Asia. J Integr Agric 13(4):660–665. https://doi.org/10.1016/S2095-3119(13)60701-7

OECD (2008) Handbook of constructing composite indicators-methodology and user guide. Organisation for Economic Cooperation and Development, Paris

Chapter 9
Managing the Nexus: The Role of Nature-Based Solutions (NBS)

1 Introduction

All three components of the nexus are closely associated with natural resources or natural capital. Strengthening the natural capital not only enhances the productivity in the food sector but also improves the resilience of agriculture and reduces its vulnerability to climate risks. Additionally, strengthening natural capital helps mitigate climate risks in the long term. The preceding analysis of various aspects of the C-N-F nexus establishes the linkages between the three components, particularly among the climate and nature components and their indicators. Though there are variations across the regions, the broad trends support the existence of the nexus. One common feature across the regions and countries is that most countries have strong proactive food policies that overcome the nexus externalities in ensuring food supplies. Across nations, irrespective of socioeconomic conditions and resource endowments, the nature component is either stable or deteriorating. This is despite the policy support for protecting and improving forest resources. Even in countries (Japan) or regions (East Asia) that are well endowed with forest resources, they have recorded a declining status of the nature component. This is mainly due to the degradation of water resources and the declining share of agricultural land. Additionally, the declining biodiversity and the quality of forest cover, which are not fully captured in our assessment, are also present across the countries. Thus, the message is clear that natural systems/natural capital need to be strengthened in order to build resilience against climate risks.

Strengthening natural capital can be achieved by reducing pressure on the natural system or through interventions that directly improve its conditions. Since the first option is a very long-term prospect, direct interventions become imperative. In this context, nature-based solutions (NBS) are a useful framework, as they address the most important societal challenges, such as food security, water security, human health, socioeconomic development, climate risks, disaster risks, ecosystem degradation, and biodiversity (IUCN 2020). NBS is defined as: "Nature-based solutions

V. R. Reddy et al., *Strengthening the Climate–Nature–Food Nexus*,
https://doi.org/10.1007/978-3-032-29423-4_9

are actions to protect, sustainably manage and restore natural and modified ecosystems in ways that address societal challenges effectively and adaptively, to provide both human well-being and biodiversity benefits" (IUCN 2020, p. 1). Nature-based solutions, in the present context, primarily involve strengthening forest resources, biodiversity, land, and soils, as well as ecosystem services such as water and air. This chapter examines the framework for nature-based solutions and the types of interventions that are part of NBS in the context of the C-N-F nexus. These interventions require financial resources as well as technological know-how; most developing countries in Asia are not in a position to manage these resources on their own. All these countries need external support. This chapter examines various options for financing the NBS interventions in developing countries in Asia.

2 NBS Framework and Practice

A practical NBS framework in the context of the C-N-F nexus is presented in Fig. 1. Forests, land, and water resources, part of the natural component or systems, are degraded, and this degradation is widespread, impacting ecosystem services and biodiversity across Asian developing countries. NBS help maximize the ability of the natural capital base to enhance ecosystem services, while also ensuring food production, climate change adaptation/mitigation, among other benefits (Matthews et al. 2019). Although NBS can be implemented from the farm level to the landscape level, it optimizes benefits at scale (watersheds, river basins, forest reserves, mangroves, etc.). NBS implementation can be location-, region-, or country-specific, incorporating local knowledge and culture, and integrated with other policies and interventions. The most important aspect of NBS is that it recognizes the trade-offs between present economic benefits and future production options and ecosystem services (Cohen et al. 2016).

NBS focuses on managing and improving the natural capital, including land, forests, and water. Management interventions could include soft management and infrastructure interventions, such as resource accounting, monitoring, and governance. Soil conservation and rejuvenation interventions, improved soil quality (biochar use), balanced input use, and crop management need both hard and soft interventions. For these interventions, carried out at the farm to landscape scale, awareness-building among farming communities is needed for farm-level adaptation. Diversifying land use, such as shifting from monoculture to multiple crops/seasons, would increase land cover for longer periods during the agricultural year, which helps mitigate nutrient depletion. These practices need to be sustained and strengthened, particularly in terms of crop selection and farm management. Adaptation of these interventions at the farm level needs awareness building and policy support.

In the case of afforestation, forest quality improvement, biodiversity protection, and improvement require both hard and soft (governance) interventions. Water, along with soil, is one of the most degraded resources that need immediate attention in terms

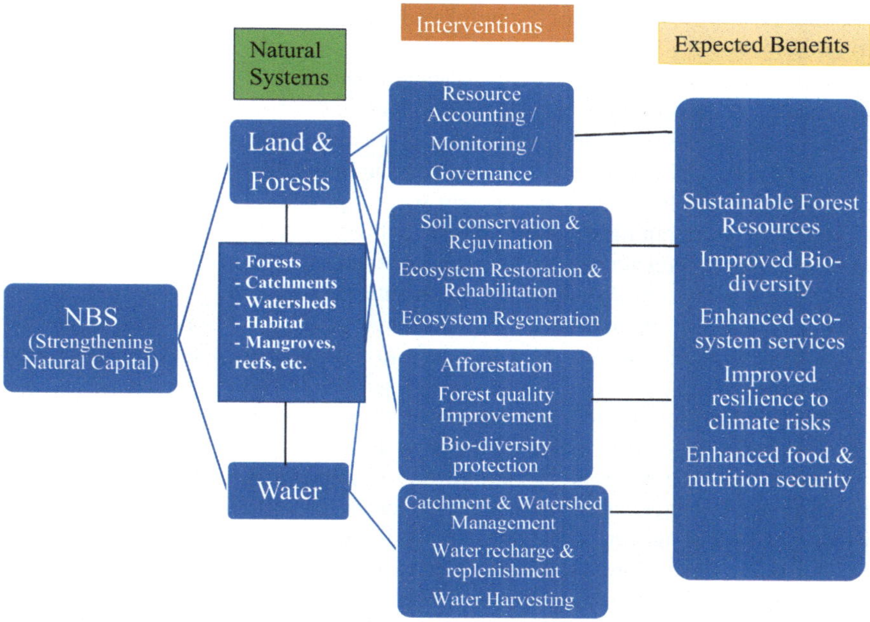

Fig. 1 Nature-Based Solutions (NBS) in the context of climate-nature-food nexus. *Source* Adopted with modification from Reddy et al. (2025)

of hard as well as soft (management) interventions. Appropriate watershed and catchment interventions serve the dual purpose of both soil and water conservation. Watershed development enhances in-situ moisture conservation, rainwater harvesting, and improved groundwater recharge. Catchment management interventions, apart from improving soil and water conservation, improve the longevity and efficiency of irrigation reservoirs and dams. Groundwater recharge, replenishment, and water harvesting interventions in urban areas are critical for supporting sustainable towns and cities. All these interventions are part of "green infrastructure" that strengthens the natural resource base, enhances productivity, improves the livelihoods of communities, and improves climate resilience (Reddy and Chiranjeevi 2020; Reddy et al. 2017).

NBS include sustainable farm practices such as biochar, nutrient management, and conservation agriculture (Iseman and Miralles-Wilhelm 2021). Increasing biochar use can enhance the carbon storage capacity of soils. The objective of nutrient management is to strike a balance between inorganic and organic fertilizers and pesticides, and to reduce the use of herbicides. Crop rotation is also a proven method for improved nutrient management and enhanced soil carbon sequestration. Systematic promotion of these interventions helps their widespread adoption and benefits (Corsi and Muminjanov 2019). For instance, promoting precision agriculture through the balanced and timely application of agrochemicals, rather than relying on organic or natural farming practices, could improve adoption rates and impacts (Reddy et al. 2019).

Multifunctionality is a basic characteristic of natural production systems. Multifunctionality is a subset of ecosystem services and is part of NBS. Multifunctionality addresses soil, water, and landscape management, enhancing ecosystem services (Reddy and Rahut 2023). Sustaining the multifunctionality of the production systems could be seen as an efficient way of production (due to the jointness of products). Maintaining the environmental flows (e-flows) is another important aspect of NBS. By considering and minimizing trade-offs, such as farm versus basin scale, production versus sustainability, and quantity versus quality, NBS can enhance or sustain e-flows in river basins. E-flows enhance ecosystem services further and ensure the sustainable livelihoods of downstream communities, apart from sustaining biodiversity.

Possible interventions include promoting appropriate cropping patterns, balanced and sustainable input use, accounting for multifunctionality, and effective water governance. These interventions require higher allocations in the form of green infrastructure, towards natural capital at different levels. NBS-related interventions, in turn, could lead to various system and individual-level benefits, ranging from ecosystem services to improved resilience, farm incomes, food and nutrition security, sustainable livelihoods, and equity. NBS have been proven to be cost-effective and help improve system resilience, and they are amenable to policy regulations and economic incentives/disincentives (Iseman and Miralles-Wilhelm 2021). Despite the significant direct and indirect benefits associated with NBS, adopting these practices is neither easy nor smooth. This is primarily due to the time lag between adaptation and the flow of benefits. In fact, due to biological processes, some practices, such as reduced chemical use, could result in production and income losses in the short run (Reddy et al. 2019).

While NBS directly improves the assimilative capacity of ecosystems, there is a need to address the increasing environmental problems associated with GHG emissions, effluents, and air pollutants. Even developed countries like Japan and South Korea are facing issues with water and air pollution. These problems require the proper identification, monitoring, and regulation of pollution sources. Addressing these problems requires effective policy design and implementation. Since these problems are mostly associated with the private sector, incentive/disincentive policies. At the same time, some of the NBS interventions, such as livestock development and the use of livestock manure, could contribute to emissions. Policy should be cautious about these externalities and ensure a balanced approach.

Besides, some of these benefits, such as ecosystem services, are social/public rather than private/individual. As a result, private investments are shied away from, and farmers are not keen to invest, as these services are not valued or compensated. Payments for environmental services (PES) rather than incentives for adopting NBS need to be adopted. At the same time, the existing incentive structures promoting unsustainable practices, viz., input subsidies, subsidized water, and power, need to be replaced with PES. This would help reduce the risk of adopting NBS. Policy makers are often wary about the resource burden and transaction costs associated with such policy interventions. Repositioning of existing incentive policies can help direct investments towards NBS (green infrastructure), as the current allocations towards NBS in total allocations are only 15% (FOLU 2019). Given the wider range of benefit flows and financing mechanisms, such as those involving the corporate sector, private–public-people modes, etc., can facilitate the implementation of NBS on a wider scale.

3 NBS Interventions: Critical Areas

In the context of the C-N-F nexus, there are several nature-based interventions that are critical for ensuring balance among the three dimensions. There are hard as well as soft interventions, as discussed earlier. While hard interventions are necessary conditions, but not sufficient to achieve sustainability in the absence of soft interventions. Hard interventions also require greater amounts of financial resources. Important hard interventions include: (i) afforestation/reforestation, (ii) catchment/ watershed development, (iii) water replenishing and harvesting, and (iv) promotion and adoption of sustainable and restorative practices. These interventions are termed "green infrastructure" that strengthen the natural resource base, enhance productivity, improve the livelihoods of communities, and improve climate resilience (Reddy and Chiranjeevi 2019; Reddy et al. 2017). And soft interventions include: (i) awareness and behavioral changes, (ii) sustainable production processes, (iii) monitoring mechanisms, and (iv) resource governance institutions. One major constraint of NB interventions is that they have long gestation periods for realizing the benefits. In what follows, we discuss the important hard interventions that require substantial financial resources.

3.1 Afforestation

Forests are often referred to as the lungs of the ecosystem. They enhance the assimilative capacity of the ecological system (absorbing GHG emissions) in addition to supporting biodiversity. Forests provide ecosystem services such as water recharge, water purification, soil conservation, and catchment protection, as well as microenvironment control. Forests can provide benefits in all three dimensions through their supportive services, such as ecosystem services (natural capital), regulatory services like climate control (climate), and provision services like food and non-food forest produce (food). The threshold level of forest cover is 33% of a country's land mass in order to maintain the ecological balance. Although Asia maintains a threshold level at the aggregate level, most countries in Asia have significantly less forest cover. The lowest is Uzbekistan, at about 10%, while large countries like China and India have less than 25% of their areas under forestry. While these countries prioritize afforestation and reforestation initiatives, they are unable to achieve their targets due to financial constraints. In fact, Uzbekistan has embarked on a massive afforestation programme with the support of the World Bank. India has been investing in afforestation for the past 3 decades with marginal success.

3.2 Catchment/Watershed Development

Catchment or watershed development is essentially a soil and water conservation intervention with long-term benefits in terms of soil rejuvenation, water recharge, and improved ecosystem services, resulting in enhanced crop productivity. They are considered one of the best technological options for degraded regions, improving natural capital and stabilizing catchments and agriculture. These interventions also act as a resilience-building mechanism against climate change (Carpenter et al. 2001) and can be used as an adaptation strategy to deal with climate change (Joslyn 2011). However, these aspects are not often integrated into the design of interventions. This is particularly critical in regions with low precipitation and water potential, which are likely to be most affected by climate variability (Reddy and Chiranjeevi 2020). Designing these interventions as climate-smart would make them effective and result in better value for money. Adoption and implementation of these interventions in Asian countries have been chequered over the decades (Reddy et al. 2017). Some countries, like India, have implemented these interventions on a massive scale for over two decades, although the policy priority has diluted in recent years.

3.3 Water Replenishment

Water is the most critical area of intervention, as all the impacts associated with nature and climate dimensions are translated through water. In most Asian countries, agriculture (including irrigation) accounts for more than 80% of the available water resources, underscoring its importance to food production. And it is one of the most mismanaged natural resources. The present analysis clearly reveals that all Asian countries face water-related problems, either in terms of quantity or quality. Even developed countries like Japan and South Korea face issues with water quality. Historically, the development of water resources has been in the form of creating huge reservoirs and dams to provide access to irrigation, drinking water, and to meet industrial needs. Over the years, these large infrastructure projects have become increasingly economically unviable and environmentally risky. Consequently, a number of Asian countries have been promoting groundwater development through private individual investments since the 1980s. Asia ranks first in the world in terms of groundwater exploitation. In the absence of proper regulation and enforcement, the exploitation and degradation of groundwater are widespread in a number of countries. Groundwater levels have been declining at a rapid pace, resulting in well failure and deteriorating water quality.

Countries like India have been reporting farmer suicides linked to well failures, and Bangladesh has been reporting life-threatening arsenic intrusion. The main reason, apart from unsustainable groundwater exploitation, is the unscientific and poor investments in replenishment and recharge mechanisms. While watershed development helps recharge groundwater, poorly designed and scaled interventions (with low coverage) have made them ineffective or yield only marginal benefits. There is a need for investments in the replenishment and recharge mechanisms. These could include environmentally less risky storage structures, rejuvenating existing and traditional water storage systems, and extending surface water supply systems to replenish water bodies and groundwater zones for recharge. These interventions need to be carried out at scale (across a river basin) in order to maintain the demand–supply balance. Nevertheless, regulating the extraction beyond sustainable yields with the help of economic and extraterritorial instruments is required to maintain the ecosystem's balance in most regions.

3.4 Sustainable and Restorative Practices

Restoration of ecosystems is a critical strategy of NBS. This should be part of the catchment or watershed development for sustainable soil and water resources at the scale of the catchment or watershed. Forest landscape restoration is another key strategy in this regard (Chaonan and Feng 2024). Depending on the status of ecosystems, the intervention strategies could involve minimal or no interventions, moderate interventions, or the creation of new ecosystems. Ecosystem rehabilitation involves actively restoring ecosystems by restoring their structure, function, and ecological processes, often through activities such as habitat reconstruction and hydrological management. (Chaonan and Feng 2024). Rehabilitation could help address climate risks through the introduction of drought-resistant tree species, carbon-absorbing species, and species to check wind and water erosion. These restoration strategies do not require hard engineering interventions that disturb the ecological balance.

Numerous sustainable farm practices are also part of NBS, including the application of biochar, nutrient management, and conservation agriculture (Iseman and Miralles-Wilhelm 2021). Increasing biochar use can enhance the carbon content of soils. The objective of nutrient management is to strike a balance between inorganic and organic fertilizers and pesticides, and to reduce the use of herbicides. Most of the countries or regions studied here do not apply chemical fertilizers in a balanced manner, i.e., excessive use of nitrogen (in the form of urea) is widespread. The use of composite fertilizer is less popular due to policy distortions (Reddy et al. 2025). Conservation agriculture, or sustainable agroecosystems, is most widely practiced across the world. Activities include minimizing soil disturbance (zero or reduced tillage), increasing soil organic cover (cultivating cover crops during fallow periods), and crop and species diversification (diverse cropping systems including tree species). Although these practices are expected to increase soil organic matter and sediment retention, they are often not adopted as a comprehensive package of practices. As a result, impacts are not clear, and adoption rates are sparse. Despite their promotion for about 2 decades, their adoption rates are quite low in the selected Asian countries. Systematic promotional activities may help facilitate their adoption in dry regions, as the practices can reduce operational costs, improve nutrient use efficiency, increase water savings, and enhance crop yield stability (Corsi and Muminjanov 2019). For instance, promoting precision agriculture through the balanced and timely application of agrochemicals, rather than relying on organic or natural farming practices, could improve adoption rates and impacts (Reddy et al. 2019). Crop rotation is also a proven method for improved nutrient management and enhanced soil carbon sequestration. All these practices help restore and improve soil quality, conserve moisture, and hence enhance the climate resilience of crop systems.

4 Financing the C-N-F Nexus

So far, the C-N-F Nexus has only been financed indirectly—in the sense that the three facets of the nexus are funded in isolation or in silos. While the cross-factor benefits of funding climate, nature, or food initiatives may be factored into the analysis of funding, they are often viewed as positive by-products rather than a focused outcome. A lack of understanding about the nexus itself has meant that funding has also failed to consider the three facets in connection with each other. Thus, climate, nature, and food all have distinct funding mechanisms, often with terms that conflict with each other's goals. There are significant benefits to packaging financing in a way that enables the development of all three facets of the nexus. It allows for more efficient allocation of capital. By enabling the targeting of multiple goals through initiatives, the efficiency of any capital allocated increases simply by virtue of achieving multiple outcomes across different areas. This efficiency can be further bolstered—initiatives must often overcome issues introduced by progress in other areas. By developing and financing initiatives that encompass all three facets, capital does not need to be allocated to undo or overcome changes or issues (externalities) caused by other initiatives.

This is most evident in the climate-food interaction. Early investment in the improvement of food production systems has resulted in adverse outcomes for the natural environment and climate. Apart from the degradation of land and water resources resulting from the excessive application of fertilizers and pesticides, activities such as crop burning significantly contribute to the release of massive amounts of carbon into the atmosphere (Lan et al. 2022). Similarly, climate change initiatives impose limits on how aggressively food production systems can be expanded and bolstered Giles. These conflicts are not limited to climate-food either. Nature-food policies also tend to conflict with each other, particularly through the channel of land use pattern regulation. Thus, initiatives that improve outcomes within individual facets of the nexus will greatly benefit from funding, as well as initiatives that treat the nexus as a package. Finally, a critical aspect of the financing pipeline that can pose a challenge in a discretized approach is the choice of funding instrument. Each of the facets has different financing pathways that best suit them. However, packaging them together may pose a challenge, but it also unlocks the potential for novel financing pathways that may prove more effective than those commonly adopted at this time.

Nature, for example, is predominantly funded through grants and trusts, known as conservation finance. Conservation finance is channeled primarily through three assets: (i) guarantees, which are a passive form of financing; (ii) debt, and finally (iii) equity, primarily in the form of junior equity as a compensation for not exploiting resources aggressively.[1] Climate is overwhelmingly funded by debt (Chiranjeevi and Rahut 2025). Ultimately, food systems are largely funded by subsidies, and local governments often offset these costs. External funding typically comes in the form of donations or grants of food grains, rather than in the form of capital or technology.

There is a difference in approach for how problems related to each of these facets are addressed. Climate issues take center stage and are generally viewed as global problems that need to be addressed through collective action, thereby attracting significant funding. A sense of great urgency marks Climate funding, and capital is more readily mobilized through far more diverse channels. Nature initiatives are often viewed as related to climate initiatives and are expected to benefit indirectly from them as well, resulting in significantly lower levels of capital availability. Thus, conservation funding is largely through charitable initiatives that primarily focus on grants as channels for mobilizing action in the space. Finally, food is predominantly seen as a problem that local governments and legislators are best equipped to address. Thus, (in comparison to the other two facets), there is limited direct international funding for food security. Financing for food security is predominantly channeled through trade agreements and subsidies. During times of significant distress, food systems are bolstered externally through charitable sharing of food grains themselves to bridge supply gaps immediately. It is thus evident that packaging all three facets of the nexus—climate, nature, and food—in order to present a cohesive funding pathway presents a unique challenge.

There is significant discourse across all three facets on the sufficiency of funding available to those seeking to solve the problems of these systems. The overall perspective is clear—the level of funding across all facets falls critically short of meeting the necessary levels. This is further exacerbated by the inefficiencies of the allocation of funding across these areas as well. In terms of packaging solutions to the combined challenges of the nexus, nature-based solutions, as discussed above, are a strong argument for the benefits of packaging these solutions together. Instead of each of these facets competing with the limited financing available, combining the pools of financing unlocks much greater potential for meeting the combined funding goals through NBS solutions. The channelization of funding for food systems through trade channels means that barriers to climate and conservation funding can be reduced by leveraging trade channels to build more comprehensive capital flows. It also unlocks access to pools of more direct funding that it has not had so far.

---

[1] (https://www.thegef.org/sites/default/files/publications/GEF%20Assembly_ConservationFinance%20Factsheet_9.4.18.pdf).

The major barriers to financing, however, continue to be at the policy level. The macroeconomics of global trade and foreign policy dictate the overall level of foreign involvement in policy and budgeting. Given that climate financing is primarily a channelization of foreign funding into developing nations, it is essential for nations receiving funding to maintain clear boundaries and caveats for the use of this funding. NBS initiatives tend to blur the lines between the three facets, which can prove to be a challenge; however, they can also unlock new opportunities. Funding can potentially be channeled through the exchange of technology, as well as other critical resources, such as rare metals and materials, required to accelerate the indigenous development of climate-mitigating or adaptive processes and technologies. These channels would help bolster all three systems, as opposed to being focused on any one facet.

Debt financing for development, although popular, is not ideal for financing public infrastructure and initiatives (Vippala et al. 2025). Debt financing introduces external pressure to finance the debt, thereby diluting the potential benefits of the infrastructure from the outset. These are challenges faced by recipients of debt financing in both the climate and public infrastructure development spaces. Sovereign debt financing is also a powerful tool in international diplomacy and foreign policy. Therefore, developing nations carefully manage their levels of foreign debt in order to protect their interests and bargaining power on the global stage. This conflict is inherently built into any foreign financing received by developing nations; therefore, any multilateral financing aimed at managing the nexus must circumvent this conflict. Multilateral Development Banks (MDBs) serve as an appropriate conduit that can help navigate these conflicts more effectively. MDBs help smooth the funding bidding process simpler in multiple ways—(i) Packaging: MDBs can help consolidate different lines of capital across the three facets of the nexus, and present a consolidated package to potential funders; (ii) Covenants of funding: MDBs can significantly help smoothen covenants of funding by better negotiating the terms of financing by acting as an intermediary between financiers and recipients—MDBs can also act as intermediary guarantors for weaker developing nations to attract more funding with less punitive terms of repayment; iii) Monitoring and Assessment: MDBs can also serve as effective intermediaries for conducting independent monitoring and assessment activities, helping to bolster funding requirement assessments, as well and helping to facilitate implementation and monitoring assessments in the recipient countries.

Thus, along with policy and research, even financing stands to benefit significantly from building a better understanding of the C-N-F nexus as a cohesive force, as opposed to discrete entities operating independently of each other. Effective financing is crucial to achieving positive outcomes for nations navigating the sustainable development landscape, and effective management of the nexus is essential to realizing long-term benefits from sustainable development activities. Effective financing and monitoring also help developing nations better align with SDG goals and unlock more financing alternatives. As can be seen, a holistic C-N-F nexus approach can potentially make a big difference in how developing nations navigate the global development financing space.

5 Conclusions

In the context of nexus, nature-based solutions appear to be effective in achieving sustainable development. Strengthening natural capital is not a policy priority in most Asian countries at the moment. The number of important nature-based interventions that help address the three dimensions of the nexus is identified. Although countries adopt some of these interventions, they are often implemented independently (e.g., land, forest, or water improvement) rather than being integrated. Financing is a complex landscape to navigate for both recipients and financiers. There are difficult decisions to be made regarding which initiatives to fund, various covenants for funding, and funding instruments. Thus far, due to the discrete approach taken towards the three facets of the C-N-F nexus, initiatives have competed with each other for financing, and in some cases, have even led to outcomes that create problems for one another. A consolidated approach is crucial to achieving positive outcomes across all facets. Nature-based solutions provide a novel approach to engaging all three facets of the nexus. Financing always catalyzes outcomes for sustainable development and climate change management. The competing dynamics of financing further complicate access to capital in these critical spaces. The nexus approach has the potential to simplify these flows, and MDBs sit at the center of this potential, unlocking capital flows and ensuring strong, positive outcomes for all participants in this space. Trade flows can also be leveraged to build more robust and dynamic forms of funding, relying on the strengths of the different predominant financing lines that climate, nature, and food systems currently enjoy to improve capital flows across the board. Thus, a progressive understanding of the nexus, paired with progressive policy and strong involvement of MDBs in negotiating and facilitating capital flows, will set the stage for strong, positive outcomes for the C-N-F nexus as a whole and for developing nations in general.

References

Carpenter S, Walker B, Anderies JM, Abel N (2001) Ecosystems 4:765–781

Chiranjeevi T, Rahut DB (2025) Smart agriculture under climate change. Asian Development Bank, Tokyo, Japan

Cohen-Shacham E, Walters G, Janzen C, Maginnis S (eds) (2016) Nature-based solutions to address global societal challenges. International Union for Conservation of Nature. https://portals.iucn.org/library/sites/library/files/documents/2016-036.pdf

Corsi S, Muminjanov H (2019) Conservation agriculture: training guide for extension agents and farmers in Eastern Europe and Central Asia. FAO. http://www.fao.org/3/i7154en/i7154en.pdf

Food and Land Use Coalition (FOLU) (2019) Growing better: ten critical transitions to transform food and land use. FOLU Coalition. https://www.foodandlandusecoalition.org/wp-content/uploads/2019/09/FOLU-GrowingBetter-GlobalReport.pdf

IUCN (2020) Guidance for using the IUCN global standard for nature-based solutions, 1st edn. International Union for Conservation of Nature Gland, Switzerland

Iseman T, Miralles-Wilhelm F (2021) Nature-based solutions in agriculture—the case and pathway for adoption. FAO and Nat Conserv. https://doi.org/10.4060/cb3141en

Joslyn O (2011) OECS Integrated Water Resource Management (IWRM) Forum St. Vincent and the grenadines PPCR/SPCR and the integrated watershed management approach, Bay Gardens, St. Lucia

Lan R, Eastham SD, Liu T et al (2022) Air quality impacts of crop residue burning in India and mitigation alternatives. Nat Commun 13:6537. https://doi.org/10.1038/s41467-022-34093-z

Matthews J, Matthews N, Simmons E, Vigerstol K (2019) Wellspring: source water resilience and climate adaptation. Nat Conserv. https://www.nature.org/content/dam/tnc/nature/en/documents/Wellspring_FULL_Report_2019.pdf

Reddy VR, Chiranjeevi T (2020) Reorienting watershed management as a climate change adaptation strategy. In: Encyclopaedia of water: science, technology, and society, John Weily

Reddy VR, Rahut DB (2023) Multi-functionality of rice production systems in Asia: a synoptic review. Asian Development Bank Institute, Tokyo, Japan. https://www.adb.org/publications/multifunctionality-of-rice-production-systems-in-asia-a-synoptic-review

Reddy VR, Rahut DB, Sonobe T (2025) Future of small holder farms in Asia. Asian Development Bank Institute, Tokyo, Japan. https://www.adb.org/sites/default/files/publication/1041036/future-smallholder-farming-asia.pdf

Reddy BG, Reddy VR, Reddy MS (2019) Potential and Constraints of Zero Budgeted Natural Farming (ZBNF): a study of Andhra Pradesh. Ind J Agric Econ 74(3)

Reddy VR, Saharawat YS, George B (2017) Watershed management in South Asia: a synoptic review. J Hydrol 551:4–13

Vippala AK, Behara B, Rahut DB (2025) Bridging the climate finance gap: unpacking the global disparities and imperatives for developing countries, China, People's Republic of, India, Philippines, Vietnam. https://www.adb.org/publications/bridging-the-climate-finance-gap-unpacking-the-global-disparities-and-imperatives-for-developing-countries

Chapter 10
Policy Insights and Future Direction

1 Background

The concept of nexuses emerged during the 2000s, with discussions on water-energy-food (W-E-F) linkages that focused on groundwater and the energy required to extract it. It received wider recognition following Hoff's research (Hoff 2011), which was subsequently expanded upon in further studies (Kurian 2017; Ratna Reddy et al. 2018). The widespread degradation of natural resources is a consequence of unsustainable use practices in recent decades. This has led to the view that the linkages of food systems extend beyond just water and energy, including biodiversity, ecosystem services, and other factors. Climate change is now further aggravating this degradation process. While a strong natural resource base would make the system more resilient to climate risks, a weak natural capital not only aggravates climate risks but also adversely impacts food systems. As the impacts of these linkages are increasingly becoming evident at the micro level, the existence of a broader nexus concept of climate-nature has been recognized in recent years. The influence of such a nexus, termed the climate-nature-food (C-N-F) nexus, is much broader and deeper, prompting policymakers and development thinkers to underscore the importance of strengthening natural capital to overcome climate risks and sustain food systems.

While the C-N-F nexus is well established at the conceptual level, there is a lack of evidence-based research examining and understanding the status of the nexus in different agro-ecological contexts. In a way, this research is among the first attempts to provide empirical evidence for establishing the presence of a nexus. Understanding and establishing the C-N-F nexus is at a nascent stage, as most research is being carried out in silos, dealing with each component separately. There are no practical case studies using real-world data, despite its potential in addressing climate risks and food security. The extent and intensity of the nexus vary across geographies as well as over time, as the three components of the nexus are dynamic in nature. The balance and management of the nexus depend on the policy and socio-economic environment of the specific region or country. This research focuses on Asia and its four

V. R. Reddy et al., *Strengthening the Climate–Nature–Food Nexus*,
https://doi.org/10.1007/978-3-032-29423-4_10

sub-regions, with a broad objective of establishing the existence of the C-N-F nexus and understanding the nature of this nexus across various agro-climatic contexts. Furthermore, the linkages, interdependencies, and trade-offs among the three components are explored. The research also examines the monitoring and management of the nexus.

2 Assessing the Nexus

The primary objective of this research is to empirically establish the existence of the nexus and quantitatively assess the linkages and interactions between the factors or indicators representing each dimension of the nexus. The direction of interactions (positive or negative) is determined according to theoretical expectations. Since we are assessing the nexus quantitatively for the first time, basic methods are adopted to establish a simple proof of concept. Higher levels of primary data analysis, utilizing more complex methodologies, will enable a more nuanced assessment and representation of the nexus. All assessments have been carried out for Asia and its four sub-regions over a time period marked by the greatest change in terms of climate, nature, food technology, and policy. Finally, an integrated monitoring index has been developed in order to assess the status of the nexus across time and space. It is essential to note that, given the preliminary and exploratory nature of the research, the present approach does not account for externalities such as industrial activity and the effects of global trade and economics. Thus, it only includes the most fundamental aspects for each of the facets, as close to independent variables as possible.

3 Insights

3.1 The Establishment of the C-N-F Nexus

The goal of this research is clearly laid out in a simple hypothesis:

> Climate, natural capital, and food supply systems are inextricably inter-linked, and are both directly and indirectly impacted by policy and activity among each of the facets. These interactions manifest in different ways, and empirical proof of these interactions is sufficient to establish the existence of the C-N-F nexus.

Through analysis of secondary data and literature, the C-N-F Interactions can be classified into three major pathways:

(1) Neglect of natural capital is aggravating climate risks and food security.
(2) Reinforcing linkages between climate and nature has a multiplier effect on the resilience of the ecological system.

(3) The extent of degradation or climate risks has yet to cross the threshold levels, coupled with the buffering of existing natural systems, the adverse impacts are often mitigated at the aggregate level.

Due to the trade-offs among the three facets, these impacts could manifest in the short to medium run, as:

(a) Declining forest cover could potentially increase agricultural land and food production.
(b) Unsustainable water exploitation (water stress) could enhance food productivity.
(c) Increased temperature anomalies could improve food systems.

Taking or considering these trade-offs at face value could prove counterintuitive in the long run and jeopardize sustainable development. The policy of "Ignorance is bliss" could prove quite problematic in the long run.

The findings from this research support the hypothesis, as well as the pathways for the interactions between the three facets of the nexus. Even the most basic analysis of secondary data provides empirical evidence to support the hypothesis and its subsequent interactions. The preliminary assessment of core indicators representing the three dimensions of the nexus over a period of three decades at the Asian level and its four sub-regions reveals clear patterns regarding the presence of the nexus. However, this does not fully corroborate with the available research based on laboratory studies or modeling (indicating clear linkages between climate and food). This does not mean that there is no C-N-F nexus. Hitherto, research has been carried out in silos, i.e., assessing one or two systems at a time, such as climate or climate-food. In the process, the linkages between nature and food systems are bypassed or ignored. Food systems heavily draw on natural systems in terms of soils, water, biodiversity, and ecosystem services. Unless the degradation of the natural system reaches a critical point (soil quality, water stress, water quality, etc.), its impact on food production or supplies will not be noticeable. More importantly, given the high policy priority of food security/self-sufficiency in all countries, technologies that promote land intensification are continually encouraged. As a result, the excessive and imbalanced use of chemicals is keeping yields artificially stable or increasing, as evidenced by the increasing levels of agricultural inputs over time (Ratna Reddy and Prudhvikar Reddy 2007). Such practices are proving to be environmentally unsustainable and economically unviable. Although the impacts are yet to be visible at the aggregate level, they are evident in some countries and regions in Asia (Ratna Reddy and Prudhvikar Reddy 2007; Ratna Reddy et al. 2025). This serves as the foundational evidence of the existence and the strong interactions between the facets of the C-N-F nexus.

The visualizations and correlation data provide empirical strength to the qualitative assessment of the interactions between the facets of the nexus. The linkages between various indicators establish a clear exploitative relationship between expanded food availability and nature degradation. Furthermore, they also highlight the acceleration of climate variability and risks, as well as the degradation of natural capital and biodiversity. Intensification also correlates with higher levels of degradation, with increased nitrogen usage linking to higher levels of forest cover loss due to the trade-off between forest and agricultural activities. This is clearly seen in both the visualizations and the correlation matrices. This analysis also corroborates the evidence of climate-nature interactions. Freshwater availability per capita is positively correlated with the percentage of forest cover. Infrastructural progress is also recognizable in the correlation values, with both freshwater withdrawals for agriculture and freshwater availability per capita being highly correlated across both the 30-year and 2010–2020 periods, with a much stronger correlation during the latter period, i.e., 2010–2020.

Closing the loop refers to the stress that climatic variance places on food production systems, contributing to higher water stress and increased water withdrawals (not to mention potential crop losses from untimely rains or flooding). Higher levels of forest cover are correlated with higher freshwater withdrawals for agricultural purposes. It is also interesting that high levels of freshwater withdrawals per capita are negatively correlated with calorie availability. This suggests that higher freshwater withdrawal and availability may also coincide with other outcomes that lead to lower calorie availability. Further, restoration of natural capital (such as expanding or preserving forest areas) may alleviate climate variability. The analysis clearly indicates that there is still potential for agriculture in the region to benefit from temperature increases (Mendelsohn 2014), as evidenced by the correlations between calorie availability and temperature anomalies, relative humidity, average annual precipitation, and precipitation anomalies. While early gains of farm technology were limited, the last decade (2011–2020) saw the accumulation of gains from continued intensification (spread to more areas) and the resulting returns from the scale application of intensification technologies and infrastructure.

Thus, within the Asia Aggregates, there is clear evidence of the Climate-Nature-Food nexus. These forces, in some cases, alleviate the negative impacts of certain variables, but largely work together to magnify the negative impacts of exploitative activities or technologies. It is especially important to note that some of the more restorative interactions between two of the factors may have detrimental impacts on the third factor in the nexus—as in the case of restorative C-N interactions leading to lower F values, higher C-F values leading to lower N values, and high F-N values resulting in lower C values. This now leaves us with the need to understand how the C-N-F nexus was handled in different regions and specific countries, given the varying prevailing circumstances, and the resulting outcomes in terms of the C-N-F factors. These aggregate findings (Asia) are, for the most part, representative of the interactions and outcomes at the country level. There are some outliers at the country level, but they are not significant enough to refute or counteract the evidence in favor of the existence of the nexus itself. Instead, the value of the outliers lies in providing context for ground realities and informing the policy and management of the C-N-F Nexus itself.

3.2 Learnings from Quantitative Assessments

A key finding across all quantitative analyses conducted in this study is that a significant amount of context and information is lost through the process of aggregation. There are significant offsetting interactions between various data points that obscure valuable insights, which can be leveraged to drive a predictive analysis. Policymakers as a whole run the risk of being drawn into a false sense of security, or rather a lack of urgency, when looking at aggregate data, which might indicate that there aren't any issues, or that the issues are emerging but not significant or critical. This is the mechanism that policymakers and financiers, for the most part, leverage as a consequence of the level of operation. It is challenging to adopt a micro-level approach to policymaking for centralized governments. Similarly, financiers can only rely on aggregate analysis presented by potential recipients, supplemented by further independent aggregate-level analysis.

Data quality and resolution have also suffered greatly as a result of this focus on aggregation. Climate data, in particular, is only widely available on a very broad geographic and temporal scale. While data on food systems is robust, the frequency with which this data is sourced has recently declined, as governments begin to consider improvements in food security and availability, resulting in lower levels of monitoring. Nature data has proven to be the most challenging to source and reliably translate into independent variables. Of the three facets, nature data had to be approximated to the greatest extent through qualitative analysis. A lack of unified work in the space has meant that the need to source data for these facets in relation to one another has been neglected thus far. However, the findings from this preliminary research make a strong case for continued work in this area and for significant investments to be made in data gathering, assessment, and policy enforcement.

3.3 Country Analysis

The assessment of the C-N-F nexus at the regional and country levels clearly indicates its presence over a longer period, as well as in recent decades. At the regional level, most countries are moving on the margins of achieving food security and self-sufficiency, although food availability/supply is keeping pace with demand. This is associated with deteriorating natural conditions or natural capital, especially in terms of water resources. Loss of biodiversity is also evident.

The context of Central Asia and Uzbekistan presents an extreme situation in terms of natural endowments, climate risks, economic strengths, and policy environment. Under these circumstances, the trends and linkages between C-N-F indicators clearly establish that the indicators of climate and nature complement each other in a way that degradation of one element can adversely influence the other. Given that Central Asia and Uzbekistan are facing natural resource degradation and climate risks, the country-level situation is significantly more difficult than the regional level situation.

While this clearly establishes a nexus between climate and nature, the joint influence of climate and nature is not being reflected in the food dimension. This is true of all the sub-regions and the case study countries. The reason is that a proactive policy environment has helped in overcoming the C-N-F nexus effectively. This is particularly evident in the context of Uzbekistan, which reported a higher total calorie supply compared to other case study countries, including Japan. While national food policies are ensuring food availability through food imports and a shift towards non-plant calories, environmental policies appear to be cognizant of the fact that continuing such support is not sustainable and needs to be addressed in a sustained manner. Keeping this in mind, Uzbekistan has embarked on a massive afforestation programme to improve biodiversity and ecosystem services.

While Nepal is in a comparable situation, its approach seems to be different. Nepal's path is to strengthen the C-N-F nexus, i.e., maintaining and enhancing economic growth at the expense of the environment and natural capital, as it can rely on its better natural resource endowments. Nepal has adopted a less intensive path for meeting its food demands. In the absence of crop intensification, forage development for livestock was given importance along with protecting its forest resources and biodiversity. Livestock and forests supported the food needs in Nepal. It appears that Nepal has managed the C-N-F nexus in a more balanced manner. This could be more due to a lack of choice rather than a deliberate strategy. Nevertheless, this is a better way of managing the C-N-F nexus when compared to the approaches of India and Bangladesh in the South Asian region. Thus, Nepal provides some important insights into managing the nexus.

As it is often argued, in the context of most developed countries, growth/ development and degradation often go hand in hand. Some exceptions to this argument are the Scandinavian countries, which are effectively balancing growth and prosperity. In the Asian context, Japan is setting such an example of effectively managing the C-N-F nexus. This could be due to the high policy priorities set for environmental protection and development, and adoption of environmentally friendly technologies across the board, including agriculture. In fact, Japan is one of the most food-insecure countries in the region, importing 60% of its calorie requirements. Although these strategies are slowing down growth and even impacting food production, they are paying dividends in the form of preserving biodiversity and ecosystem services. These strategies range from promoting sustainable farm practices and payments for ecosystem services to farmers to the promotion of green and sustainable technologies. While the number of countries in Asia has similar policies, most of them remain on paper, and very few are implemented. However, Japan could implement all these policies on the ground, and the results are evident. Japan provides important insights into the C-N-F nexus and learnings for systematically managing the C-N-F.

The case of Southeast Asia and Vietnam provides another interesting scenario for the C-N-F transition over the period that has evolved alongside economic transformation from agriculture to manufacturing and then to services. Unlike in the case of East and South Asia, there is no overbearing or dominant nation in the region, as all the member countries are small in size and demographics. Despite the homogeneity in size, resource endowments, climate risks, and the trajectory of economic transformation among the small nations of the region, they differ widely in their well-being, i.e., the extent of poverty, food security, GDP growth, etc.

Vietnam is a better-performing nation in the region, with consistent growth rates, food security, and a low incidence of poverty. This is achieved through a singular focus on food security through intensification and industrial growth, with little concern for biodiversity. While this has helped achieve food security and economic growth, it has come at the cost of natural systems, resulting in biodiversity loss and declining ecosystem services. Such a siloed approach is not sustainable in the long run. Unless a comprehensive and integrated approach to C-N-F is adopted, it is more likely to result in inefficient and undesirable outcomes. Vietnam offers valuable insights into why a linear approach, focusing on either of the three dimensions (C-N-F), is ineffective in addressing sustainable development.

3.4 Indexation

Integrated indices help determine the status of the C-N-F nexus across sub-regions and countries between two time periods. Indices clearly indicate that most sub-regions and countries are on the margins (close to zero) in terms of status, although they are in the negative in the majority of cases. There are variations across the sub-regions and countries. It is observed that most sub-regions and countries have focused on food security or the food component to the neglect of the nature component. Thus, the indices indicate that the C-N-F Nexus is not moving in a balanced or desired direction, which may impact the resilience of the natural system. While focusing on food security helps address the immediate priorities of hunger and poverty, the imbalance between nature and food could have an adverse impact on food systems in the medium to long term. Neglect of the natural systems would aggravate the pace and intensity of climate risks. Such adverse impacts are yet to be conspicuous at the aggregate level to prompt policy action and monetary support. Most countries seem to be focusing on quick fixes for addressing climate risks rather than devising comprehensive approaches to build and strengthen resilience through investments and financing of natural capital or biodiversity initiatives. Indexation channelized the analysis on the nature of the indicators and how they translate into interactions within the C-N-F Nexus. The preliminary work of analyzing the direct indicators through secondary analysis has helped establish the most fundamental data required to analyze the nexus, which can be leveraged to conduct primary data analysis at a more granular level.

4 Policy Insights and Future Direction

- Ensuring food security has been at the center of policymaking for most of these countries so far. It is clear that food security achieved at the expense of long-term natural systems is inherently unsustainable. In this context, the experiences of nations like Japan (high income) and Nepal (low income) provide crucial insights into how a well-managed C-N-F Nexus is possible, regardless of the economic strength (or weakness) of a country.
- The cases of Uzbekistan and Vietnam serve as cautionary tales. While Uzbekistan, is at the early stages of benefitting from the adoption of agricultural expansion (in terms of stabilizing of harsh climate, and improvement in forest cover), it runs the risk of over-indexing on this activity beyond a threshold level, resulting in deteriorating outcomes over a longer time-frame, as evidenced in the case of Vietnam, which is now struggling with severe degradation of natural resources, despite starting from a position of natural wealth (in terms of forest cover and water availability).
- Policy in Uzbekistan also recognizes the importance of boosting natural capital. However, the strong reliance on a non-plant-based caloric supply chain already shows an inherent conflict between food security and ecological conservation policies for the country. An inability to assess the C-N-F Nexus as a cohesive system will tend to result in conflicting policies and, therefore, unfavorable outcomes for developing nations.
- Most countries seem to be focusing on quick fixes for addressing climate risks, rather than devising comprehensive approaches to build and strengthen resilience through investments and financing of initiatives that fortify natural capital and other biodiversity conservation initiatives.
- Nature-based solutions present a robust case as a pathway for effective nexus management. Despite gaining prominence in recent times, they still lack the strong policy focus required to ensure broad-spectrum adoption.
- NBS struggles with long gestation periods that hurt adoption rates. These gestation periods also tend to diminish financiers' willingness to fund such initiatives, as risk and cost scale with time. Innovations that reduce the long gestation periods of NBS could prove to be an effective catalyst in improving adoption rates at all scales.
- MDBs stand to play a critical role in helping make NBS and other novel solutions the standard approach for sustainable development. Currently, funding for the various facets of the C-N-F Nexus remains fragmented. Policy and financing for the three facets either compete with each other or must differentiate themselves sufficiently to attract more diversified capital flows.

- MDBs can bridge this gap in understanding (through funding research) as well as finance. They are uniquely positioned to package the nexus as a cohesive problem and bring efficiencies in terms of capital management, as well as initiative/intervention design, execution, and monitoring. They can also unlock previously inaccessible capital flows, which would greatly benefit the developmental and conservation efforts required to manage the nexus effectively.

Paving the way for the future

- Policymakers rely predominantly on aggregate-level analyses, which can result in errors in decision-making. However, at the aggregate level, many of the ground-level challenges and issues tend to be smoothed out, as demonstrated in this study. The outcomes for climate and development policies in recent years can be attributed to this lack of visibility of issues resulting from this approach. As the old saying goes, the devil is in the details. Unless policymakers delve into the details, they risk remaining disconnected from the real problems faced by the people.
- Furthermore, by the time micro-level issues snowball to the aggregate level in terms of visibility, the cost—both in terms of effort and resources—may already be beyond the means of developing nations, and may be too late to solve altogether, regardless of global collaboration.
- At the aggregate level, if challenges are seemingly being conquered, there is poor economic incentive for financing programs that enable micro-level assessment. Losing access to this granular data runs the risk of unintentionally impeding related research. This is quite visible from the gaps in data observed during this study.
- Conducting micro-studies to identify emerging challenges and vulnerabilities related to the nexus and its cross-cutting impact is critical for designing better-informed policies.

References

Hoff H (2011) Understanding the nexus, background paper for the Bonn 2011 nexus conference: the water, energy and food security nexus solutions for the green economy. Stockholm Environment Institute, Stockholm

Kurian M (2017) The water-energy-food nexus: trade-offs, thresholds and trans-disciplinary approaches to sustainable development. Environ Sci Policy 68:97–106. https://doi.org/10.1016/j.envsci.2016.11.006

Mendelsohn R (2014) The impact of climate change on agriculture in Asia. J Integr Agric 13(4):660–665. https://doi.org/10.1016/S2095-3119(13)60701-7

Reddy VR, Reddy PP (2007) Increasing costs in agriculture: agrarian crisis and rural labour in India. Ind J Labour Econ 50(2), April–June

Reddy VR et al (2018) A water–energy–food nexus perspective on the challenge of eutrophication. Water 10:101. https://doi.org/10.3390/w10020101

Reddy VR, Rahut DB, Sonobe T (2025) Future of small holder farms in Asia. Asian Development Bank Institute, Tokyo, Japan. https://www.adb.org/sites/default/files/publication/1041036/future-smallholder-farming-asia.pdf

The manufacturer's authorised representative in the EU is Springer Nature Customer Service Centre GmbH, Europaplatz 3, 69115 Heidelberg, Germany. If you have any concerns regarding our products, please contact ProductSafety@springernature.com

Printed and bound by CPI Group (UK) Ltd, Croydon, CR0 4YY
17/09/2026
02218458-0002